heartland

FRONT COVER
Oliver Freeman, Christopher Main
and Alex Freeman on Stringybark Hill,
Treetops, Cootamundra district,
about 1980.

OPPOSITE
Egret in Flight. Sculpture of mild
steel by Michael Murphy, 1998.
Installed above a roundabout in
Peter Street, Wagga.

April Sharman planting old man
saltbush and acacias on Oakvale,
Narrandera district, August 2001.
Courtesy Danceplant.

Floodwaters sweeping Wallendoon
Street, Cootamundra, December 1919.
Courtesy National Library of Australia.

George Main grew up on the southwest
slopes of New South Wales and now
lives nearby in Canberra on the southern
tablelands. He is employed as a curator
in the People and the Environment
Program at the National Museum of
Australia, and is involved in several
regeneration projects in the
Cootamundra district.

heartland
The regeneration of rural place

GEORGE MAIN

A UNSW PRESS BOOK

Published by
University of New South Wales Press Ltd
University of New South Wales
Sydney NSW 2052
AUSTRALIA
www.unswpress.com.au

© George Main 2005
First published 2005

This book is copyright. Apart from any fair dealing for the purpose of private study, research, criticism or review, as permitted under the Copyright Act, no part may be reproduced by any process without written permission. Inquiries should be addressed to the publisher.

National Library of Australia
Cataloguing-in-Publication entry

Main, George V., 1970– .
Heartland: the regeneration of rural place.

Includes index.
ISBN 0 86840 873 5.

1. Land use, Rural – New South Wales – Cootamundra Region. 2. Nature - Effect of human beings on – New South Wales – Cootamundra Region. 3. Human ecology – New South Wales – Cootamundra Region. 4. Human settlements – New South Wales – Cootamundra Region. 5. Wiradjuri (Australian people) – History. 6. Cootamundra Region (N.S.W.) – History. I. Title.

994.48

Design Di Quick

contents

acknowledgments	viii
introduction	1
mastery	16
elsewhere	57
progress	88
division	121
silence	161
revolt	200
regeneration	228
epilogue	251
notes	262
index	281

For Indigofera

acknowledgments

Land and people led me to write this book. In the Cootamundra district, my father taught me to sense the rich human histories of places. My mother instilled a love for plants of the southwest slopes, and first introduced me to a chocolate lily. The brave and creative responses of Neil Murray to the history and life of the country that raised him inspired and motivated me.

I would like to thank my academic supervisors – Libby Robin, Debbie Rose and Peter Read – for guiding my work with such care and generosity. Other staff and many students at the Centre for Resource and Environmental Studies and elsewhere at the Australian National University offered help and encouragement.

Over several years, farmers Graham Strong and Owen Whitaker gave deep insights into mainstream and alternative styles of agriculture. Alec Hansen shared understandings of local ecologies gathered over decades of close observation. Wiradjuri Language Development Project teachers Stan Grant and John Rudder patiently conveyed aspects of Wiradjuri language and culture. Gudhamangdhuray elder Bob Glanville readily gave advice and time, and shared family stories.

Many other people offered time and thoughts. I am indebted to Diana James for convincing me to send chapters away to a publishing house, and to publisher Phillipa McGuinness for her interest and enthusiasm. Deepest thanks to Penny Taylor for giving support and ideas during an extremely busy time in our lives.

acknowledgments (ix)

I am grateful to Land and Water Australia for funding my research and for organising meetings with like-minded scholars. Many thanks also to National Museum of Australia staff for granting me time to work on the manuscript and for general support. Image costs were supported by ARC Discovery Grant DP0208361. Russell Brooks skilfully prepared the index.

Thanks to Mushroom Music Publishing for permission to publish lyrics sung by Jimmy Little, to Margaret Connolly and Associates for permission to publish Les Murray's poem 'Sydney and the Bush', and to Kevin Gilbert Estate for permission to publish extracts from *The cherry pickers* by Kevin Gilbert.

introduction

Streetlights bled into winter fog as we drove into town to catch the night train, dark paddocks and stony hills behind us. Kids returning to boarding school, heavy with sleep and woollen jumpers, hauled suitcases into carriages from a platform of bitumen and decorative cast iron. Rolling away from Cootamundra, the train climbed the southwest slopes of New South Wales towards the southern highlands and Sydney. Outside, a rising moon followed behind a tracery of old paddock trees, frost gathering on fence posts and crop land – cold country stripped and pushed. Morning sun warmed us as the train passed urban granaries and flourmills. Inside Central Station, pigeons flapped through cavernous spaces as we trudged past commuters to catch suburban trains to boarding houses.

Back then, the pathway we took from the country to the city seemed vital somehow, a nourishing relationship. Today, Sydney appears oriented elsewhere, towards global networks of power and commerce, away from inland places, from rural terrains tired and faltering.

Almost two decades after leaving the Cootamundra district for boarding school and university, questions rise from the wheat and sheep country of my childhood, from a hillside rippling in midday summer heat.

Only an elderly yellow box tree breaks the golden expanse of wheat, halfway down the gentle slope. Eucalypt and acacia saplings grow in the distance, where the wheat crop meets a fence.

(2) *heartland*

The revegetation planting forms a long band of dark green. Crop dust billows behind humming machinery nearby, as a bulky harvester sweeps the contour. Good rain fell over autumn and winter. Inorganic fertilisers generated a heavy crop of grain. Herbicides and insecticides ensured that only humans would reap the produce and nutrients of the paddock.

The vast blanket of wheat is part of what is often termed our 'ecological footprint', the cost imposed on nature to maintain human bodies. Problematic assumptions underlie the notion of human footprints on nature. Modern people tend to see themselves as divorced from land and natural systems. Meeting our physical needs, we assume, inevitably burdens local ecologies.

Does human presence unavoidably bring destruction? Are we not of the earthly nature our bodies sense, materially embedded in dynamic and intricate patterns of life? Do we really not belong here, as constructive members of vibrant biological communities? Could it be possible to instead imagine and enable the flourishing of living systems and people as one?

Muttama Creek emerges from dense forests of ironbark and Cootamundra wattle on the watershed dividing the Murrumbidgee and Lachlan river basins. The stream curves southeast through paddocks and the town of Cootamundra, beneath road and railway bridges, and joins the Murrumbidgee River upstream from Gundagai.

The southwest slopes tumble and subside towards the western Riverina plains from the mountains and tablelands of the Great Dividing Range. As in most parts of Australia, plants native to the region are adapted to fire and drought. In autumn and winter, unless El Niño prevails, moist air drifts eastwards and rises with the land. Gentle rain swells creeks and rivers. Fertile slopes turn green. On summer days, the surging calls of cicadas fill the bright air. Grasshoppers leap from dry grass.

The Cootamundra District Hospital opened on the northern bank of Muttama Creek in 1909. Almost eight decades later it closed, as the state government 'rationalised' country hospital services. Behind the old hospital, a cement path leads to a mown area where the maternity building used to stand.

Spring sunshine drenched the district and town of Cootamundra the afternoon I visited my birthplace. I lay down on a sparse lawn of kikuyu and flowering capeweed, and imagined my mother following the path one cold August night more than thirty years ago, painful contractions gripping her body. Pages of my notebook turned in the warm breeze. Pigeons flapped and cooed, perched on the rusted iron roof. Afternoon sun warmed the length of my body. I watched small black ants move through grass, across compressed earth. Occasionally, I lifted my head to take notes. Here, my lungs first drew breath – air damp above Muttama Creek, shifting through trees, over paddocks and streets, my skin sensing movement. South and west, at the edges of town, rounded granite hills formed a close, curving horizon. As I lay in sunlight, on the ground near the waterway, I felt secure, held by the land.

'Cootamundra is the centre of one of the most fertile farming and grazing districts of the Riverina', *The Land* newspaper declared in 1970, the year I was born beside Muttama Creek. In the district lived

> a most progressive group of farmers and graziers ever anxious to heed the advice of Department of Agriculture experts and to implement new techniques in pastures, crops and animal husbandry.[1]

Today on the southwest slopes, farmers and scientists face an array of ecological disorders and challenges induced by an industrial, export-oriented model of agriculture. Drought, bushfire and frost haphazardly undermine production and profitability, and worsening problems of dryland salinisation and soil acidification emphasise the ecological instability and vulnerability of modern farming systems. A team of environmental scientists recently placed Muttama Creek in the 'high environmental stress' category. Across the entire Murrumbidgee River catchment, only Jugiong Creek, a waterway draining the Harden and Binalong districts, received a higher ranking of ecological sickness.²

As afternoon shadows lengthened beside Muttama Creek, I walked away from the hospital building and crossed a bridge. At a street corner, I noticed stormwater drainpipes emptying a trickle of water into a ditch leading to the creek. Cumbungi grew where the pipes opened onto a bed of silt.

The wide pipes drain the southwestern side of town – a sunken area, formerly swampland, now a matrix of streets, houses, sports fields, a primary school. Drainpipes and other engineering works ensure the expanse below the granite hills rarely floods.

Brochures issued by the Cootamundra Development Corporation to attract visitors and business investment do not dwell upon the lost swampland or the Wiradjuri origins and cultural context of the placename 'Cootamundra'. Beside photos of grain silos, railway lines and wide crops, glossy publications describe a 'progressive' town, 'a prosperous valley', 'high agricultural activity' and 'commercial and industrial growth'.

'Cootamundra', we learned as children, meant something like 'turtle in the swamp'. The placename 'is derived from an Aboriginal word *Gooramundra* or *Goodamundry*', records a local promotional booklet published in 1972, 'the meaning of which is given variously as *turtles*, *marsh* or *swamp*, *low lying*.' According to recent linguistic research, the original placename was 'Gudhamangdhuray', a Wiradjuri name for an area of Muttama Creek swampland and for the local clan. 'Gudhamang' is a species of freshwater turtle, possibly the eastern snake-necked turtle, and the suffix '–dhuray' means 'having' or 'with'. People of the Cootamundra area perhaps called themselves 'Gudhamangdhuraymayiny', Wiradjuri elder and language teacher Stan Grant explained. The added suffix '–mayiny' means 'people', forming a word denoting 'turtle-having-people'. Gudhamangdhuray clanspeople probably considered freshwater turtles kin, family to care for and receive life from, descended from the same ancestral Dreaming figure as themselves.³

According to explorer and anthropologist Alfred Howitt, the 'Kutamundra' clan was a major Wiradjuri group between the southern tablelands in the east and the western Riverina plains. Howitt recorded the meaning of 'Kuta-mundra' as 'river turtle'. The swampy place on Muttama Creek where colonists established the town of Cootamundra, he noted, was the heart of Gudhamangdhuray clan territory.[4]

Gudhamangdhuray clanspeople saw freshwater turtles as significant, ecologically, socially and spiritually. From a Wiradjuri perspective, no fundamental divide exists between human culture and animal nature. Local people took responsibility for nurturing turtle populations in Muttama Creek swampland. In return, freshwater turtles gave life and identity to people. Gudhamangdhuray clan elder Bob Glanville told me his ancestors reserved the swampy place as a turtle sanctuary. Hunting was forbidden there by law, allowing populations of the reptiles to flourish undisturbed. In good seasons, they would spread beyond sanctuary boundaries into areas where turtle hunting was permitted. During droughts and other disruptive natural events, sanctuary law ensured the turtles survived to repopulate local waterways.

Mary Gilmore grew up southwest of Cootamundra at Brucedale, near Wagga. In the winter of 1878 she began her working life assisting her uncle, George Gray, a Cootamundra schoolteacher. The writer and poet explained how Wiradjuri applied sanctuary laws to protect and nurture animals and plants. Places reciprocated the protection and life people gave to other species and local ecologies:

> All billabongs, rivers and marshes were treated as food reserves and supply depots by the natives. The bird whose name was given to a place bred there unmolested. The same with plants and animals. Thus storage never failed.[5]

When settlers arrived on the southwest slopes, they described grassy woodland and swampy places teeming with diverse life. Through sanctuary regulation and other strategies designed to promote ecological connectivity and wellbeing, Wiradjuri fostered dense and varied populations of plants and animals. Biological diversity ensured abundance of food and materials and bolstered the resilience of land.

Like those of other Aboriginal groups, Wiradjuri systems of tending local ecologies incorporated social and spiritual concerns. An intricate and holistic cultural framework shaped actions towards other species and the land.

British colonists imported a different set of attitudes and beliefs to rural Australia, a complex system of knowledge forged in the industrial and scientific revolutions of western Europe and infused with ancient Christian and agricultural traditions. As railways extended inland from Sydney late in the nineteenth century, a modern, industrial style of farming spread across hillsides and creek flats. 'Even amongst agriculturalists there are a large number that have not yet realised the fact that they belong to an industrial occupation that stands far away ahead of any industrial occupation as regards the scope there is for the utilisation of scientific knowledge', the *Agricultural Gazette of New South Wales* declared in 1898.[6] Settlers applied scientific methods and industrial technologies to harness land for export-oriented production. In the nineteenth century, argues historian Heather Goodall, the inland slopes and plains of Australia were no 'pre-modern rural pastorale'. Rural colonisation represented

> the modern itself, in its relentless application of new technologies to the landscape, its rapid embrace of 'labour-saving' innovations and the continuing expectation that engineering approaches will solve resource problems.[7]

In southern Australia, on the western slopes of the Great Dividing Range, settlers perceived biologically diverse living systems of grassy woodland and swamp as chaotic and disordered, as wild terrain requiring subjugation and cultivation. In reality, the complex and interconnected natural patterns tended by Wiradjuri people were organised and effective. 'The city desk of a newspaper, a rabbit's intestines, or the interior of an aircraft engine may look messy', writes political scientist James Scott, 'but each one reflects, sometimes brilliantly, an order related to the function it performs.'[8] Physicist and philosopher Fritjof Capra describes a shifting order 'manifest in the richness, diversity and beauty of life all around us.' Ecological dynamism enlivens and shapes the complex patterns of nature. Biological diversity ensures natural stability and productivity. 'Throughout the living world', Capra observes, 'chaos is transformed into order.'[9] In rural Australia, European colonists imposed a foreign landscape ideal of regular simplicity. The dispossession of Aboriginal clans, the drainage of swamps, the erasure of diverse communities of grassy woodland life and the establishment of crop monocultures and pastures comprising few species disordered and destabilised local ecologies.

In the first decades of agricultural development on the southwest slopes, removal of native vegetation and repeated cultivation degraded soil structures and exposed land to the erosive powers of wind and water. Throughout the region, heavy rainfall carved deep gullies into hillsides, jagged emblems of instability and disorder. Late in the twentieth century, new technologies and chemical farming systems reduced the need for cultivation and enabled restoration of soil structures. Severe erosion events are now rare on the southwest slopes. Fertilisers, herbicides and insecticides often enable spectacular crop yields. Towns bustle with economic activity.

Unfortunately, the dominant, industrial style of agriculture weakens and destabilises living systems. Modern agriculture relies primarily on linear flows of external inputs, not ecological connectivity and nutrient cycles. Fragmentation of dynamic natural patterns makes farmland vulnerable. Since the late nineteenth century, the widespread erasure of interrelated plant, animal and insect communities has destroyed the biological diversity and ecological connectivity required for land to be strong and naturally productive. Storms, frosts, cold winds, droughts, fires, insects and other natural forces regularly impose disaster across wide, exposed paddocks.

Efforts to address ecological disorders in Australian farming regions usually take place within the practical domains of environmental science and natural resource management. Explorations of the historical and cultural dynamics responsible for the realities of rural Australia are rare.

The ecological problems we face today across the globe originate 'in human behaviour, in complex socioeconomic practices with long histories', argue scholars of western culture Jill Ker Conway, Kenneth Keniston and Leo Marx.[10] To be effectively addressed, they must be placed within the elaborate historical and cultural contexts from which they arise.

Across time, culture shapes human behaviour. 'We treat people, places and things in accordance with the way we perceive them', observes farmer and writer Wendell Berry.[11] On the southwest slopes of New South Wales a uniform aesthetic of crops and pastures, towns and homesteads, fences and railway lines, and common patterns of ecological disorder suggest a shared perceptual framework. Established beliefs and attitudes block the emergence of alternative ways of seeing and engaging with land. 'Most of the constraints working against environmental change are cultural: we have to know ourselves as well as the country', writes environmental historian Tom Griffiths.[12]

The methods and technologies of modern agriculture are only the external aspects of an industrial system of primary production. 'It also has an inner dimension', historian Donald Worster notes, 'vast, complex and effective: the habits of thought and perception that are needed to make the system and its demands appear reasonable.'[13] What dynamics of imagination and history transformed the grassy woodlands and swamps of Wiradjuri country into a modern agricultural region? What particular habits of thought and perception delivered dryland salinisation, soil erosion and acidification, dying paddock trees, local extinctions? What cultural processes maintain the dominant model of industrial agriculture? Might alternative styles of imagining and engaging with rural places return ecological wellbeing and natural productivity to the agricultural heartlands of Australia?

In the chapters that follow, I explore the intricate cultural and historical context of ecological change and disorder across the southwest slopes of New South Wales. Patterns of settler engagement with places, with Wiradjuri people and with other species reveal much about beliefs and values underlying and shaping rural Australia. As settlers bound the western slopes to global trading networks in the nineteenth century, the relentless and blind demands of distant markets began reworking the land. Faith in linear notions of progress justified dramatic ecological and social change. Understandings of people as fundamentally separate from living systems enabled efforts to erase and silence the life of the land.

Inevitably, disruption and loss brought disorder and rebellion. Resistance shown by land and people to the monological power of modern industrialism offers hope. Despite the widespread erasure and silencing of dynamic and diverse biological communities, actions based on intimacy with the life and particularities of places are today bringing regeneration.

~ ~ ~

The colour photo in the album of yellowing pages is square with rounded corners, a product of the 1970s. My sister leans on a decorative cast iron fence around an old grave, her face grim. My brother, a toddler, pauses as he climbs the ironwork set in stone. The camera caught only half of me. I stand at the edge of the image, my right hand on the fence. Behind us, afternoon light casts the shadow of a crucifix across the roof of a small church. Why were we standing by the grave of our grandfather's grandmother, looking so solemn? 'Sacred to the memory of Maria, the beloved wife of T. H. Mate', the headstone reads behind the iron fence, 'Died June 23 1876, Aged 60 years.'

Maria Bardwell and Thomas Hodges Mate sailed to Sydney on the *Palambam* from England as free settlers in 1833. They met on board, and later married.

Thomas and Maria Mate prospered in Australia. They established a squatting run at Tarcutta, east of Wagga. The main Sydney to Melbourne road, later the Hume Highway, ran through the fertile area held by the newcomers. They took advantage of the increasingly busy route, opening a hotel and store beside Tarcutta Creek, opposite their station homestead. Thomas, a conservative, represented the Hume electorate in the Legislative Assembly of New South Wales throughout the 1860s. He fiercely opposed political efforts to allow closer settlement and agricultural development across vast pastoral estates like his own.

A biography of the squatter and politician appeared in *Australian men of mark*. Published in 1889 to celebrate a century of Australian colonisation, the leather-bound, gold-embossed volume contains life histories of influential male settlers. Commentators deeply valued the vigorous colonising work of Thomas and Maria Mate. 'I am sure that the Tarcutta people', wrote early Riverina colonist James Gormly, 'are proud of those very worthy pioneers Mr. and Mrs. T.H. Mate and their family.'[14]

Our visit to the Mate family graves at Tarcutta in the middle of the 1970s extended the honouring of British 'pioneers' and the dramatic changes they imposed. Beside the old church, we learned patrilineal identities infused with notions of colonial triumph and status.

Thomas Mate headed inland soon after arriving in Australia from England. He worked on Cunningham Plains, near Murrumburrah, to gain skills in pastoralism and station management. In 1836, one account holds, Mate travelled southwest from the Murrumburrah district in search of strayed cattle.[15] He eventually found them near Tarcutta Creek, on the other side of the

Murrumbidgee River far to the southwest, and decided to establish a squatting run there.

Could the lost cattle have travelled so far and crossed the Murrumbidgee, a major waterway? Thomas Mate, it appears, made the most of his mission to find them, venturing further to secure land for himself and his family. Reports of 'depredations' by Aborigines beyond Tarcutta, noted Mate's biographer in *Australian men of mark*, stopped him exploring for land south of Tarcutta Creek. According to the entry, the new squatter avoided conflict with the many Wiradjuri people of the Tarcutta district:

> His method was kindness and firmness. He insisted on the aboriginals obeying his orders, and he faithfully kept his promises to them. The result was that, though he has had three or four hundred camping round his place at one time, they never killed or even molested a single person on the station, or did any appreciable damage during a period of forty years. On the contrary, they were made very serviceable at lambing and shearing times, and were otherwise useful about the station. It is some years now since the last of the aboriginals disappeared from the neighbourhood. When the townships were formed at Wagga Wagga and the surrounding centres, the blacks were gradually attracted thereto, and by degrees died away by the use of spirits and other results of their contact with civilisation.

Perhaps Mate and local Wiradjuri didn't clash violently. Clanspeople undoubtedly knew the potential outcomes of resisting the 'firmness' and defying the 'orders' of the newcomer.

Elsewhere on the southwest slopes, violence did erupt as tense relations between squatters and Wiradjuri collapsed. Outside the Wiradjuri Regional Aboriginal Land Council office in Wagga, west of Tarcutta, a brass plaque titled 'Statement of the Wiradjuri Nation' honours Wiradjuri opposition to colonisation, 'when the rivers ran with the blood of our ancestors leaving the name Wiradjuri embedded forever in Australia's history.'

Pastoral and agricultural expansion required the erasure of

Aboriginal people and the taking of productive country. Land Council manager Roly Williams told me he knew no Wiradjuri families connected by ancestry to Tarcutta. Turbulent histories broke ties binding Wiradjuri to places. Roly talked about the forced movement of his people onto Aboriginal reserves like Brungle, in the foothills of the Snowy Mountains, Warangesda, on the Murrumbidgee downstream from Narrandera, and Hollywood at Yass, where the southern tablelands rise. When the New South Wales parliament granted the Aborigines Protection Board powers to compulsorily remove children from families in 1916, another wave of dislocation and disruption struck the Wiradjuri community.

There are meagre accounts of individuals who were, perhaps, the last remaining Tarcutta district Wiradjuri. Elderly local residents remember Tang, an Aboriginal man who worked as a vegetable gardener for Alfred Mate, son of Thomas and Maria. Tang grew watermelons so large they were carted in wheelbarrows, Olive Parramore told me. According to Tarcutta historians Bill and Fay Belling, the old man loved cricket and fishing, and died in a Wagga hospital soon after Alfred Mate and his family moved to Sydney in 1918. Nobody I spoke to knew if Tang had a different name or any descendants.

Chum White talked of another old Aboriginal man, Ned Turner, who also worked for the Mate family. When Chum was a boy, before World War I, Turner gave him a silk handkerchief for his birthday. Chum recited meanings of local placenames he suspects Ned Turner taught him. 'Tarcutta', Chum learned, is the apostlebird, the spoken name echoing the call of the species.

Across the Tarcutta district, a rich store of local ecological understandings and cultural heritage vanished as local Wiradjuri died or joined kin on distant missions and reserves.

In 1839, three years after Thomas and Maria Mate came to Tarcutta Creek, Lady Jane Franklin, wife of Sir John Franklin, governor of Van Diemen's Land, embarked on a journey from

Melbourne to Sydney. Halfway, at Tarcutta, Franklin and her companions camped near swampland on Tarcutta Creek. As darkness descended, Franklin gazed through trees towards fires burning in the Aboriginal camp. Inside a hut of eucalypt slabs she met a young man, Thomas Mate's brother. The youth told her 'the native name for this place was Umumby', she wrote in her diary.[16] Thomas and Maria Mate called their pastoral station 'Umutbee', a local Wiradjuri name for the Tarcutta Creek swampland. Next morning, Aboriginal camp residents threw boomerangs for Franklin. She met a man called Daptoe, an elder wearing a copper breastplate engraved with emus and kangaroos. 'Natives have a name for every creek every hill', Franklin noted at Tarcutta.

Beside the Melbourne to Sydney road, Umutbee Swamp featured in *Bailliere's New South Wales gazetteer and road guide*, carried by travellers in the 1860s and 1870s. Just upstream from Umutbee village Tarcutta Creek widened into a swampy expanse, thirteen kilometres long. Umutbee Swamp became a shallow lake in wet seasons, the *Guide* explained, 'visited by innumerable waterfowl.' Harry Podmore, my grandfather's cousin, worked as a butcher in Tarcutta. He remembers swans flying over the town to nest on swampland upstream. About thirty years ago, Harry told me, landholders used new, powerful earthmoving machines to drain the remaining parts of Umutbee Swamp. Cattle prices were high, and the dark, rich swamp earth grew dense pasture.

Landscape ecologists Justin Nancarrow, Robert Cawley and Tim Smith recently studied the dramatic changes a long history of pastoralism and agriculture has imposed on Tarcutta Creek.[17] Instead of the complex system of pools and swamps encountered by British explorers, the scientists found an eroding channel curving through the valley, delivering salty water and silt into the Murrumbidgee River. Few native plants remained. Vigorous weed populations and unrestrained access by sheep and cattle threatened to further degrade the stream.

Facing such ecological instability and loss, the scientists asked, 'Where do we start and what do we do?'

In light of the disorder British colonisation brought to the continent, the question may be recast as a dilemma for every settler Australian. Instead of perpetuating historical patterns of ecological and social injustice, can we think and act in healing ways? In 'space that is fragmented, in places that are broken, in the knowledge that much damage is not even visible, only certain kinds of actions are ethically possible', suggests anthropologist Deborah Rose.[18] Rose draws on the writings of Emil Fackenheim, a philosopher who advocates a process of 'turning toward' damaged and subjugated people to seek dialogical, respectful relations. Rose extends the concept to embrace wounded places and other species. Turning towards others and starting dialogue brings recognition of shared embeddedness in both natural systems and the present moment. The careful process offers hope 'that rupture and wounds are not the only forms of action we will ever produce.' As Rose points out, there are risks involved when we turn hopefully towards people, other beings, and places. In a society taken by linear notions of progress and monological relations of power over nature, we cannot know what to expect when we stop to listen and feel.

Returning to Tarcutta recently, I drove along Mates Gully Road, beside the steep, eroding banks of Mates Gully Creek. Fast food shops and petrol stations line the Hume Highway through the village. I turned off the relentlessly busy highway and headed uphill. Inside the old brick church, a stained glass window memorialising Thomas Mate shows grain-filled heads of wheat and grapevine roots entwined in rich soil. Outside, amid the Mate family graves, I encountered glossy, purple blossoms of chocolate lilies swaying on wiry stalks. The scent of fine chocolate filled the warm spring air. Wiradjuri people in the Tarcutta district probably harvested and roasted the watery, edible tubers of the lily, a common food plant of southeast Australia.

Chocolate lilies became rare on the western slopes of the Great Dividing Range as farming systems intensified after World War II. Crops and 'improved' pastures replaced grassy woodland remnants. Standing on the hillside beside the graves, I heard trucks rumbling into Tarcutta on their long journeys between cities. I knelt on damp red earth to draw in the dreamy aroma of a chocolate lily blossom. Ants crawled over rotting leaves. One deep breath, then another, as the flowering lily drew me into relationship with particularities and histories, into dialogue with place.

mastery

Securing the land

On the main street of Stockinbingal, opposite the railway line and wheat silos, two small buildings house the local museum. Inside one, a timber box sealed by glass sits on a table. The box contains a diorama of a horrific event played out by carved stone models. Five Aboriginal warriors attack a hut of eucalypt slabs and bark. Two of the men protect themselves with wooden shields as they launch spears from woomeras. One spear has struck the front door. Another is embedded in the slab wall of the hut. A dog strains on a chain, leaping at an Aboriginal man climbing through a side window. A maimed or dead European woman lies outside, her hair spread to one side, a spear protruding from her stomach. The front windows of the hut are boarded. Perhaps children cower inside. Glued as decorations onto a frame around the frightful scene are specimens of plants native to the southwest slopes: black cypress pine seedpods, bracken fern, drooping she-oak cones and chunks of lichen peeled from stone.

The word 'Wiradjuri' has two parts. 'Wiray' means 'no', while the suffix '–juri', or perhaps more accurately '–dhuray', denotes 'having' or 'with'. The Wiradjuri language belongs to a group of similar languages spoken across inland New South Wales. Each language in the group has a name beginning with a prefix meaning 'no' and ending with a suffix meaning 'with'. North of Wiradjuri country, for example, Wayilwan people say 'wayil' for 'no'. In Wayilwan, the suffix '–wan' denotes 'with'.

Wiradjuri academic Norm Sheehan told me why his people and neighbouring groups defined themselves as 'no-with' or 'no-having'. If people wanted to enter an area or gather resources there, they had to seek permission from those with legal rights over the place in question. Within the framework of Wiradjuri law, Sheehan explained, certain individuals and clans held rights and responsibilities in relation to particular areas, and thereby had power of veto over requests made concerning those areas and the resources found there. Wiradjuri legal rights extended across the greater part of three river basins west of the Great Dividing Range: the Wambuul, Galari and Marrambidya, known generally today as the Macquarie, Lachlan and Murrumbidgee.

Sarah Musgrave was born in 1830 at Burrangong, near the watershed of the Lachlan and Murrumbidgee rivers, where colonists later built the town of Young. During the first years of colonisation, her memoirs reveal, settlers acknowledged and worked within Wiradjuri systems of law.[1] James White, Musgrave's uncle, held Burrangong pastoral station. Establishing a station in the early days 'was a simple enough matter', Sarah Musgrave explained, 'after one had reached a desirable place on which to squat, providing, of course, one got a passport from the blacks'.

James White travelled well outside the area administered by colonial officials when he came to Burrangong in 1826. Vastly outnumbered by Wiradjuri, and without recourse to colonial military or police, White negotiated with a local lawman for approval to stay in Wiradjuri country. Sarah Musgrave recounted the delicate negotiations that took place when her uncle set up camp for the first time. A senior Wiradjuri man approached White:

> At first the chief disputed with Mr. White the possession of the land, but, under the influence of many gifts from Mr. White's stores, the black chief became friendly, and allowed the embryo squatter to remain, guaranteeing him immunity of attack from the tribe.

Musgrave's uncle called the Wiradjuri man 'Cobborn Jackie', and gave him an inscribed brass breastplate, 'the permanent emblem of Jackie's office as the king of the tribe'. According to historian Heather Goodall, the presentation by squatters of 'king plates' to senior clansmen reflected understanding and acknowledgment of Aboriginal law, 'where the elderly individuals who held greatest authority were those who spoke for the land.'[2] At Burrangong, after processes of exchange and negotiation, the senior Wiradjuri lawman allowed White to establish his squatting run.

Cobborn Jackie probably knew what possibilities lay in resistance. Only three years before, colonial officials had declared martial law to suppress a violent uprising of northern Wiradjuri clans around Bathurst. Although the numbers killed were never recorded, historian Peter Read suggests that hundreds of Wiradjuri people died in the conflict, perhaps one third of the local population.[3] These events may have encouraged the lawman of Burrangong and other elders of southern Wiradjuri clans to approach colonists with caution and seek agreeable relations.

Observance of Wiradjuri law by some squatters failed to maintain peace on the southwest slopes. In the early decades of colonisation, relations between Wiradjuri and settlers were fearful and unstable. Explorer Charles Sturt camped south of Burrangong squatting run, near Gundagai beside the Murrumbidgee River, in November 1829. Aborigines were 'hostilely inclined' on one station established a few months previously, Sturt noted in his journal. He worried for unarmed stockmen there 'wholly without the means of defence'.[4]

As pastoral settlement intensified, it appears that force replaced negotiation as the primary method of securing grazing land. Governor Bourke legalised the occupation of inland parts of New South Wales in 1836.[5] In the Murrumbidgee squatting district, resident Commissioner of Crown Lands Henry Bingham wrote three years later that 'arms are required in order that the Servants may be enabled to protect themselves from the Blacks'.[6]

Many feared the possibility of Wiradjuri fighting with guns. Bingham suggested that the governor of New South Wales order settlers to 'not in future give the Natives arms or ammunition, on any account'.[7] Squatting licences would not be renewed, Colonial Secretary Edward Deas Thompson announced, 'of any Persons who may give Arms or Ammunition to the Blacks, or whose Servants may be found to have supplied them with such articles'.[8] Thompson urged squatting district commissioners to coax guns from Aborigines already armed.

Male colonists sought control over the fearful situation. James Gormly came to the Riverina in the 1840s. Male settlers usually arrived alone in the early years, Gormly explained, as 'the presence of hundreds of the aborigines along the Murrumbidgee and Tumut rivers made it unsafe for women and children to be left alone in the remote inland districts.'[9] Henry Cosby was the first commissioner of crown lands for the squatting district between the Murrumbidgee and Lachlan rivers. In 1839, he estimated there were only fifty or sixty women in the local population of about two thousand mostly convict and ex-convict settlers.[10]

Men dominated Australian rural society. Mary Gilmore grew up in the Wagga region in the 1860s and 1870s, and remembered a childhood 'in a world of men'. While domestic duties kept most women indoors, men 'made the world' on outback roads, mustering cattle, breaking horses and making endless talk 'of a kind that was full of the colour of life'.[11] Her father and other settlers worried about Wiradjuri procuring guns:

> So the poor black had to die before arms became his possession; before women had to be shut in a room fearing the firing of a roof, fearing the failure of their men's limited ammunition, and having for their only comfort the knowledge that, when the time came, enough bullets would be saved for each of them and their children: as I knew they would be saved for my mother and me.[12]

While some historians question the precise accuracy of Gilmore's accounts, they agree her writings hold a general truthfulness.

Conflict she describes between settlers and Wiradjuri on the southwest slopes during her childhood, for example, probably took place decades before, or perhaps further inland where frontier dynamics still prevailed. On the inland slopes and plains of Australia, 'much human life has been sacrificed to the manes of sheep or cattle', wrote explorer and surveyor Thomas Mitchell in 1848.[13] Settlers and domestic livestock arrived in numbers on the southwest slopes of New South Wales more than a decade before colonial officials established a presence in the region.

Vague references suggest bloody encounters. Mat Sawyer's grandfather 'fought the blacks' to establish Eulomo, an extensive pastoral station near Bethungra, the grazier noted in 1937.[14] Placenames chosen during the squatting period allude to atrocities. Eulomo station was once part of Ironbong squatting run. Colonial records made in the 1840s name an eroding waterway on the boundary of Ironbong 'Slaughter House Gully'.[15] Near Temora, settlers called a rise 'Killing on the Pinnacle'.[16] Did these placenames record violence committed against Wiradjuri people, or the erasure of other impediments to colonisation? Feral horses, for example, were a great nuisance to early pastoralists on the southwest slopes. Killing on the Pinnacle and Slaughter House Gully are not inscribed on maps of the Temora and Bethungra districts published today. Perhaps only the land remembers what events unfolded at those places where sheep now graze and crops grow.

Southwest of Temora and Bethungra, Poisoned Waterholes Creek curves through paddocks below the Murrumbidgee River near Narrandera. Brutal conflict in the Narrandera district is well documented. In February 1839 on Yonco station, a stockman searching for a horse noticed crows circling. He approached the site beneath the birds, expecting to discover a dead beast. Instead, the stockman found the body of missing convict John Williams, a spear wound in his back.[17] Throughout 1839, Wiradjuri fighters forced settlers to abandon stations across a wide stretch of country downstream from Wagga.

In the early nineteenth century, historian Bill Gammage explains, the area around the Murrumbidgee River west of Wagga belonged to the Ngarrangdhuray, a major southern Wiradjuri clan based around Narrandera.[18] With the help of allies from other regions, Ngarrangdhuray clansmen slaughtered cattle and stockmen in a sustained and coordinated effort to regain country. Berembed station alone lost one thousand cattle.[19] 'The Settlers on the River are in a Great state of alarm', wrote Commissioner Cosby in May 1839,

> not daring to go out even the shortest distance from their huts, except in parties of two or three, well armed, & they are obliged to desert their lower stations altogether, finding it impossible to persuade any men to remain at them.[20]

A Narrandera district squatter told James Gormly how Ngarrangdhuray killed cattle:

> The plan the black men adopted was to hide in the reeds that grew on the water's edge, and watch until the cattle went down the steep bank to drink. Then the blacks would range themselves along the top of the bank and spear the cattle that were below them.[21]

Colonial officials, Gammage suspects, chose to ignore the horrific methods taken by squatters to regain control. Frank Jenkins came to the district in the early 1830s to graze cattle south of Narrandera with his father and brother.[22] As an old man, Jenkins told James Baylis that when Ngarrangdhuray clanspeople forced squatters to abandon their Murrumbidgee River pastoral stations,

> all the settlers on both sides of the river determined to give them a lesson; so one day they all went out armed and drove the blacks before them, who took refuge on an island thickly overgrown with reeds in the middle of the river, about seven miles up from the town of Narrandera, and here they were shot down in numbers. The island is known as the Murdering Island to this day.[23]

Not every local squatter joined in the killing of men, women and children. At the time of the conflict, James Devlin held Ganmain and Deepwater stations, upstream from Narrandera. According to one of his descendants, Devlin abandoned his stations 'owing to the blacks spearing his cattle.'[24] Despite his stock losses, the squatter respected the actions of the Ngarrangdhuray:

> He would not shoot the blacks himself nor allow any of his men to do so. He said these people were human beings like ourselves, and were only doing what they themselves would probably do under similar circumstances. Mr. Devlin took up some country near Yass at Blackall Range and moved the cattle from Ganmain and Deepwater there.

Devlin returned to his Murrumbidgee stations when the trouble passed. The squatter rebuilt relations with surviving Ngarrangdhuray: 'He used to shoot a bullock regularly for the blacks so they would have plenty of food'.[25] In the 1860s, the Devlin family gave an inscribed brass breastplate to Peter, a Wiradjuri elder on Ganmain station.[26] Donated to the Wagga Wagga and District Historical Society by a member of the Devlin family a century later, the breastplate reads 'Peter, Ganmain, Murrumbidgee'.[27]

Wiradjuri survivors of violence and disease maintained distinct lifeways during the squatting period. Squatters did not intensively manage their wide pastoral stations. They ran sheep and cattle beside swamps and major waterways, usually with the help of some convict and free workers. Extensive pastoralism did not, it appears, dramatically change the ecological character of grassy woodlands. Stock roamed distant parts of squatting runs only when intermittent creeks and waterholes held enough drinking water. In the 1870s, Harry Kavanagh spent part of his childhood at Sebastopol, a gold mining area between Junee and Temora. He remembered Aboriginal men and women living in the district, a local group of ten. The men worked on stations as shepherds. Some of the Aborigines sold settlers sheets of bark stripped from eucalypts.[28]

Five to six hundred Wiradjuri people lived along and between the Murrumbidgee and Lachlan rivers in 1844, Edgar Beckham estimated. Groups lived relatively independently of squatters:

> The Natives have no fixed places of residence, although each tribe have their own particular country and water, which they Seldom leave except for the purpose of waging war or to celebrate some jubilee with a neighbouring tribe. I have generally found the Blacks wandering within what they term their own ground, and most frequently forming their camps in the vicinity of the Settlers' Huts.[29]

Naturalist and travel writer George Bennett visited Murrumbidgee River grazing properties near Jugiong and Gundagai in 1832. He observed Wiradjuri hunting, fishing and gathering. On Darbylara station, near where the Tumut River joins the Murrumbidgee, Bennett saw Wiradjuri living as they chose:

> There were a number of the aborigines about this farm, who made themselves occasionally useful by grinding wheat, and other occupations; but no dependence can be placed upon their industry for they work when they please, and remain idle when they like; the latter being of the most frequent occurrence; but they are encouraged for their valuable assistance in finding strayed cattle, as they track the beasts with an accuracy seldom or never attained by a European.[30]

During the period of relatively peaceful coexistence between settlers and Wiradjuri, pastoralists retained the upper hand. Bound to colonial and global markets, Wiradjuri country now served demands rising in distant places for wool and meat.

Powerful landholders erased obstructions to pastoral activities. In her poem 'The Hunter of the Black', Mary Gilmore described the activities of a man hired by Riverina squatters to track down and murder troublesome Wiradjuri individuals and groups.[31] As a child, she heard of the killer working on Mimosa station, southwest of Temora, where he tracked and shot an Aboriginal man who had taken a sheep. The unnamed hired hand, Gilmore claimed, slaughtered perhaps a thousand clanspeople:

> Tomahawk in belt, as only adults needed shot,
> No man knew how many notches totalled up his lot;
> But old stockmen striking tallies, rough and ready made,
> Reckoned on at least a thousand, naming camps decayed.

Wiradjuri faced new challenges in the 1860s and 1870s as intensive farming began to replace extensive grazing across the southwest slopes.

In 1861, the Legislative Council of New South Wales approved a number of bills designed to loosen the exclusive grip of squatters over land. The legislation, collectively called the Robertson Land Acts after reformist politician John Robertson, did not apply to inland districts until 1866.[32] The same year, Donald and Mary Cameron moved from the southern tablelands to the Riverina with baby Mary, their first child. Donald Cameron worked as a stockman and builder on pastoral stations across the region. With a growing family, Donald and Mary eventually settled near Brucedale north of Wagga in 1874.

Donald Cameron deeply respected Wiradjuri people and their knowledge. Senior clansmen showed him how they tended local ecologies. Cameron and his daughter Mary were close. Listening to her father, and engaging with Wiradjuri herself, she grasped some of the intricacies of Wiradjuri ecological understandings and practices. Later in life, as Mary Gilmore, she recorded detailed memories of her Riverina childhood.

Mary Gilmore's reminiscences describe a dramatic turning point in the ecological and social history of the southwest slopes. For decades before the Cameron family came to the Riverina, Wiradjuri lived with a degree of freedom alongside pastoralists on extensive grazing stations. In the 1860s and 1870s, the closer settlement legislation and the arrival of the Great Southern Railway brought profound change. Town growth and agricultural development displaced Wiradjuri from station camps and waterways. Clanspeople could no longer tend and maintain the complex, lively patterns of grassy woodlands and swamps. With sadness, the Cameron family saw agricultural development impose loss and disorder.

Like other Aboriginal groups across Australia, Wiradjuri clans reserved places where no hunting, fishing, gathering or burning was allowed.³³ The sites held special religious and social significance. Animals and plants flourished inside the sacred refuges, spreading beyond sanctuary boundaries to replenish populations legally available for hunting and gathering. According to Gilmore, Wiradjuri reserved Parkan Pregan lagoon on the Murrumbidgee floodplain at North Wagga for pelicans, swans and cranes. Westward, on Deepwater and Ganmain stations, Ngarrangdhuray clanspeople tended swan and duck sanctuaries:

> The law of sanctuary in regard to large or wide breeding-grounds, such as Ganmain and Deepwater, where once there were miles and miles of swamps (as also down near Deniliquin), was that each year a part of the area could be hunted or fished, but not the same part two seasons in succession.³⁴

Sanctuary regulations fostered vast populations of various species. Often, as a child, Mary Gilmore heard thunder in a cloudless sky. She remembered running terrified to her mother:

> And she would tell me it was swans in the distance beating their wings as they readied for flight. Later on I learned to recognise the sound, and to listen to it unafraid.³⁵

Graziers thought immense flocks of swans nesting at Wiradjuri sanctuaries a nuisance. Reeds polluted by the birds repelled cattle from drinking places. As livestock ate feathers trapped in grass, feather-balls gathered inside their stomachs, eventually killing them. Concentrated populations of swans, Gilmore noted, enriched the soil and naturally boosted its productivity. Squatters did not recognise or value the ecological offerings of the swans, and rejected Wiradjuri sanctuary regulations in brutal style. Gilmore wrote of the 'swan-hoppers'.

> Their work was to hop the swans off the nests in the breeding-season, and smash the eggs. It was filthy work; they reeked of the half-hatched and the addled, and their trousers grew stiffer and stiffer, and filthier and filthier, as the yolks and the whites of the

smashed eggs set in the material of which they were made. The old cattle town of Wagga Wagga once had its swan-hoppers on all the stations round about; and the more they stank the prouder they were.[36]

Wiradjuri tended open eucalypt woodlands of relatively moist, soft earth cloaked in kangaroo grass, wallaby grasses and snow grass. Chocolate lilies, leopard orchids, yam daisies and other forbs with edible tubers flourished amid grass tussocks, across spaces regularly opened by flame. On one occasion during her childhood, as ground cooled after the passage of fire, Gilmore watched the 'extensive planting of seed'. Wiradjuri women gathered seed from shrubs and grasses. Gilmore helped collect grass seed, which then was rubbed and shaken in a bark container to separate grain from husks. 'The separated seed was sorted', she explained,

> the unsound or small rejected and the best planted in the burnt area but very lightly covered. It was not scattered, it was put in in small pinches so that if some failed there would still be enough for a tussock.[37]

Once a Wiradjuri woman scolded Gilmore for spitting out the seed of an unusually large 'ground-berry'. The woman searched the grass to find the seed, 'so that it could be put back where it originally came from and a strong plant grow from it.'

Wiradjuri planted seeds of quandong, small trees with pale, narrow leaves and large red fruit. Near Bethungra, Gilmore watched Wiradjuri pollinate quandong blossoms with flowering branches carried from another grove. Elsewhere, she saw branches left under quandong trees to indicate that the pollinating work had been done and didn't need repeating.[38] As with the selection and planting of large ground-berries and grass seeds, Wiradjuri used cross-pollination techniques to favour highly productive quandong strains.

A cousin of Mary Gilmore's wore a necklace of particularly large quandong stones harvested from trees on Malebo Hill, north of Wagga. The trees, Gilmore noted, 'were among those specially crossed and bred' by Wiradjuri.[39]

Mary Gilmore remembered birds hopping along ant trails after rain showers, beaking water pooled on the bare surfaces. Soil compacted by the passage of countless generations of ants held shallow threads of water for short periods. In dry country, ant trails became rare temporary reservoirs. Wiradjuri taught Gilmore and her siblings to preserve ants and pathways trodden by the tiny insects. Grass tussocks grew bigger and stayed green longer beside ant tracks. Moisture seeped down, watering deep roots.[40]

On the southwest slopes, colonists encountered living systems tended and shaped by people. Wiradjuri lived as agents of ecological connectivity, enabling a nourishing and stable web of relations. Subtle action informed by intimate understanding kept land strong and naturally productive. Ecological stability and resilience secured Wiradjuri against drought and other dramatic natural events.

In the final decades of the nineteenth century, as farmland spread and settlement intensified, Wiradjuri could no longer enforce sanctuary law or maintain established ways of engaging with country. The Cameron family witnessed a decline in the natural productivity and bounty of land and river systems. Even though fewer people now lived beside the Murrumbidgee River, fish and freshwater lobsters became scarce.[41] Settlers engaged with local ecologies differently from Wiradjuri. Sometimes deliberately and sometimes unwittingly, the newcomers restricted rather than promoted diversity and abundance:

> I do not remember in just what year it was, but the chief of the tribe at Wagga Wagga in talking to my father, said that, white settlement increasing along the river, it was not only fished in by the settlers, but fished in season and out, so that the breeding-stocks were diminishing as well as the grown fish which the blacks' laws allowed them to take for sustenance.[42]

When Gilmore first knew Parkan Pregan lagoon beside the Murrumbidgee at North Wagga, it was simply covered with pelicans, teal, duck, cranes, and swans; but being specially a pelican sanctuary, these birds predominated. When I first went to the Wagga Wagga school, as we trudged in from Brucedale Road, where I remembered clouds of them there were seventy only, then forty, then twenty, then four, and then there were no pelicans at all. The swans went till there were but two; the ducks came only at night – the few that survived.[43]

Wiradjuri complained with bitterness to Donald Cameron about the destruction of native animal populations by settlers.[44] Cameron listened and acted. He argued for the maintenance of Wiradjuri sanctuaries on Deepwater and Ganmain stations, 'to be held as such in perpetuity for the people.'

Cameron and several other Wagga men tried to enforce the wide boundaries of an emu sanctuary on Eunonyhareenyha station, northeast of Wagga.[45] The Wiradjuri placename 'Eunonyhareenyha', according to Gilmore, meant 'the breeding place of the emus'. For a short while, the men convinced people not to shoot emus on Eunonyhareenyha, or to hunt there with dogs at nesting time. When Cameron counted the once numerous emu flock inside the sanctuary, only a few hundred birds remained. 'Then', wrote Gilmore,

> the town growing, and land-settlement increasing, there was objection made that one of the sweetest spots for grazing should be set aside for birds, when selectors could farm and make homes there.

Department of Lands officials opened to selection the part of Eunonyhareenyha 'semi-reserved' for emus. Cameron spoke with the station manager, who then erected notices banning shooting and dogs. 'Unluckily the eggs were forgotten', wrote Gilmore of the action to protect the emus,

> so next year when we drove out to see them there were only about half a dozen flocks of young birds to be found in the whole area. The nests had been raided everywhere.

Cameron made other attempts to reserve land for wildlife. His daughter remembered him returning home excited one evening, 'saying that the larks were coming back again.'[46] On a grassy flat beside Houlaghans Creek, northwest of Wagga, Cameron counted a hundred groundlark nests. Flocks of groundlarks nesting among tussocks had been thinning and disappearing, as agricultural development erased and modified grassy woodland. Gilmore recalled how the brown, mottled birds shot into the air when disturbed, and 'glittered like sparks in the sun, as they mounted and sang in their myriads.' Her father built a log fence around the creek flat to exclude horses and cattle. Grass tussocks thickened, sheltering the nesting larks. Travellers on a passing road noticed the dense grasses, and put horses inside the enclosure to graze. Cameron found the nests trampled, the air above empty and silent. The event grieved him:

> After that father went by a different road to town. He had loved the larks, and they were gone. As to the fence, it became a neighbour's firewood.

~ ~ ~

Between Wallendbeen and Stockinbingal, Congou Hill rises from a patchwork of crops and pasture. Bush cloaks the southern and western sides of the tall, stony feature. One spring afternoon I turned off the Stockinbingal road and parked under the shade of yellow box trees. An open paddock lies between the roadway and the dark forest on the hillside. I opened and closed a wide gate and walked with my dog along a farm track crossing the paddock. Startled ewes and lambs bleated on the other side of the fence. I noticed a house nearby, over the creek, nestled inside a garden. No one had driven on the trail for some time – twin sandy ribbons unmarked by tyres, groomed by wind and rain.

The slope steepened. We left the track and entered the shade of the bush. Yam daisies and waxlip orchids flowered beneath red stringybark and white box trees. At the top we rested on a jumble of boulders, behind a screen of hickory wattle and black cypress. My dog snapped at flies.

The stone cairn of a trig station stands on the highest point of Congou Hill. A squared timber post holds interlocked iron disks aloft, punctured by shotgun pellets and scratched with graffiti.

I gazed north and east, across green rectangles of pasture and golden expanses of flowering canola.

Early surveyors marked Congou Hill on the boundary of Wallendbeen squatting run.[47] Troubled relations between Wiradjuri and settlers threatened the viability of grazing on parts of the southwest slopes when Alexander Mackay came to manage Wallendbeen for an absentee owner in the early 1840s.[48] Throughout 1843, the *Sydney Morning Herald* reported, Aborigines 'slaughtered, speared, or carried off far into the bush' almost an entire herd of six hundred cattle on Bogolong station, in steep country southeast of Wallendbeen near the Murrumbidgee River.[49]

Squatters gave Wiradjuri provisions, 'which tends much towards keeping the natives from Spearing and driving off the Cattle from their runs', local Commissioner of Crown Lands Edgar Beckham noted in 1844, 'and consequently prevents any disputes or collisions between the Settlers and Aborigines.'[50]

According to Sarah Musgrave, Mackay presented a brass breastplate to a senior Wiradjuri man, 'King Congo of Wallendbeen'.[51] Perhaps the gift helped consolidate a carefully negotiated relationship initiated by Mackay to ensure his cattle and stockmen remained safe from spears.

Around Wallendbeen and Stockinbingal, records suggest, relations between settlers and Wiradjuri had been turbulent for years. In the early 1830s, Ned Ryan based himself at Galong to the east, and was the first to graze stock around Wallendbeen.[52] 'Ryan had a lot of trouble with the blacks', wrote Frank Clune after speaking with Donald Mackay, Alexander's son. Ryan abandoned his western holdings around Cootamundra and Wallendbeen.[53]

Absentee pastoralist John Hurley established Cootamundra squatting run late in the 1830s. His workers grazed stock in 1839 around Congou Creek, beside Congou Hill, but were forced to retreat 'owing to the depredations of the blacks', explained Kenneth Mackay, Donald's brother.[54]

Land through which Congou Creek flows is particularly productive. Alexander Mackay quarrelled with neighbouring squatters for exclusive rights to graze stock there. He thought one parcel of land beside Congou Hill 'certainly worth contending for as it was capable of depasturing 4000 sheep'.[55]

Mackay was a 'grand old pioneer', fellow colonist James Gormly remembered, 'who reclaimed many stations from the wilderness'.[56] Perhaps the man known as King Congo of Wallendbeen, to whom Mackay reportedly gave the breastplate, belonged to the Wiradjuri clan that was responsible for evicting Ryan and Hurley.

Exchange and negotiation were probably not the only methods squatters used to secure the grassy slopes of the Stockinbingal and Wallendbeen districts for grazing. Congou Creek flows northwest and enters Bland Creek downstream from Stockinbingal. A reporter journeyed down the Bland in 1879. Up to four hundred Wiradjuri once gathered beside a waterhole on Morangarell station for ceremonies, the journalist learned:

> Of this large aboriginal population there remains but one survivor, old black Peggy, who clings to her ancestral home with a tenacity which nothing but death will sever.[57]

Mary Morton grew up on a farm near Grogan, on the Bland between Stockinbingal and Morangarell. From her nursing home window in Temora, she cherishes the view towards the Narraburra Hills she knew as a child. Inside the home, the elderly woman told me of one early landholder in the Grogan district known for 'hunting the Aborigines'.

Steel Caldwell, a Morangarell pastoralist, also had a dark reputation. 'For many years, it is said', wrote historian Veronica

McNamara, 'he kept a photo-lithograph of himself and a number of men hanging two blackfellows from the limb of a tree: The offence ... butchering a sheep.'[58] Locals told McNamara of atrocities committed alongside Burrangong Creek, a waterway flowing into the Bland downstream from Morangarell:

> The natives were very plentiful in this fertile neck of the woods, and the cruelty they suffered from the hands of the settlers in that area was appalling. One hears stories of setting strychnine baits – when the death was too long, and lingering. Arsenic was put in the treacle, which the natives were very fond of. It is said: that after going around one trail, the dead counted, amounted to two hundred.[59]

From Congou Hill I gazed northwest towards Bland Creek, Grogan and Morangarell. Afternoon sun cast shadows of eucalypt trunks over the loose forest floor as I climbed down the slope. Beyond the bush, I put my dog on the lead and rejoined the sandy path leading to the public road. With a mild jolt I noticed fresh tyre tracks. My father had warned me about walking in paddocks with the dog: 'If they see her running about, chasing sheep, they'll shoot her.' It seemed that someone did notice us crossing the open paddock earlier, perhaps from the nearby house.

Reaching the road near my parked car I opened and then pushed shut the farm gate, made sure the chain was fastened, and walked away from the grassy paddock, fenced and secure.

Fighting timber

Alec Hansen suggested we meet early at Jindalee State Forest, on the quartz and shale ridge dividing the Murrumbidgee and Lachlan river basins. I parked beside ironbarks and Cootamundra wattles. Alec was already there, under the trees, peering at something, thinking. Leaves and bark crunched as we lifted ourselves over the fence and walked up the stony slope.

Alec told me these forested hills once looked very different. In the late 1940s, he worked in the kitchen of the Silver Star café in

Cootamundra, owned by his cousins. About once a month, two retired farmers arrived by train from Stockinbingal and lunched at the café. From an early age, Alec had taken a deep interest in natural history and in the bush. He asked the elderly men what the country had looked like when they were young. There was no forest across the Jindalee hillsides in the 1860s, the old farmers told him. Stock grazed grassy expanses peppered with ancient eucalypts. Some of the old trees had trunks eight feet wide.

Alec knew a place in the forest where, years before, workers had felled a huge tree with a crosscut saw. The old stump was much wider than the trunks of trees growing there now. After speaking with the elderly men, Alec returned to the stump at Jindalee. He stepped from it out in different directions and discovered remains and impressions of other giant eucalypts. Signs of vanished trees lay roughly a hundred yards apart. In some cases, half a stump remained. Elsewhere he found broad depressions scattered with charcoal and ringed with earth uplifted by roots.

According to historian Eric Rolls, Aborigines on the northwest slopes of New South Wales regularly burned patches along grassy creek flats. Hunters speared kangaroos and emus drawn to the sweet regrowth rising from burnt ground.[60] Missionary James Günther recorded Wiradjuri words associated with land burning practices. 'Bimbarra' meant 'to set the grass on fire', while 'bimbai' signified 'a spot where the grass has been burnt.'[61] In 1848, explorer Thomas Mitchell described how Aborigines used fire to maintain open grassy woodlands across inland slopes and plains:

> Fire, grass, kangaroos, and human inhabitants, seem all dependent on each other for existence in Australia; for any one of these being wanting, the others could no longer continue. Fire is necessary to burn the grass, and form those open forests, in which we find the large forest-kangaroo; the native applies that fire to the grass at certain seasons, in order that a young green crop may subsequently spring up, and so attract and enable him to kill or take the kangaroo with nets. In summer, the burning of long grass also discloses vermin, birds' nests, &c., on which the females and children, who chiefly burn

the grass, feed. But for this simple process, the Australian woods had probably contained as thick a jungle as those of New Zealand or America, instead of the open forests in which the white men now find grass for their cattle, to the exclusion of the kangaroo, which is well-known to forsake all those parts of the colony where cattle run.[62]

Open grassland with scattered trees and shrubs, as remembered at Jindalee and reimagined by Alec Hansen in the 1940s, cloaked most of the southwest slopes when colonisation began. In the middle decades of the nineteenth century, surveyors and government officials often made observations like 'Good Grassy Open Forest', 'Extensive Flat, Thinly Timbered', 'Good open forest (Box, Gum, Wattle)' and 'Open country suitable for agriculture'.

More trees grew as the introduction of domestic livestock, the ending of Wiradjuri burning practices and the extinction and decline of native animals induced chaotic ecological responses. James Gormly came to the Riverina in the 1840s. 'I have been fighting timber all my life', he told a forestry inquiry at Wagga in 1908.[63] Gormly found yellow box trees the hardest to kill. The eucalypt regrew swiftly, and some patches needed repeated clearing. Absence of fire gave seedlings a chance to grow.

Wagga land agent Leonard Fosbery described to inquiry commissioners the vigorous responses of plant communities across places extensively ringbarked in the 1870s. Intense stocking had denuded the properties of grass. Fire could no longer sweep the bare paddocks, Fosbery explained, and germinating white cypress pine grew unchecked.

> Only give the forests an opportunity and they will re-afforest themselves. I have seen land which had been cultivated for thirty years, and when let run to grass there has been millions of seedlings shoot up as thick as wheat ... I have had country in the Riverina where the timber growth has absolutely starved me out, and I have been forced to abandon the country because I could not keep the timber down.

Rising forests of eucalypts and other woody plants unnerved settlers. In 1881, a group of pastoralists on the southwest slopes petitioned the Legislative Assembly of New South Wales. Responding to the spread of white cypress pine, they asked for help in 'exterminating this powerful enemy of the State and its subjects'.[64] Early colonists of the Temora district 'faced unknown and constant peril', wrote local historian Rob Webster in 1950,

> not only from the aborigines, who had to be skillfully handled, nor from the shattering change of season from drought to flood which a young and unprepared country must face, but also from the unremitting imprisonment and separation caused by the thick and dangerous bush which surrounded them for mile upon mile on every side.[65]

Settlers used the technique of ringbarking to kill trees. With a sharp axe they cut a ring of sapwood from each trunk. Above the wound, where sap could no longer rise, the tree wilted and died.

Resilient plants, eucalypts usually produce fresh shoots beneath debarked rings. In 1908 George Sutton, manager of the Government Experimental Farm near the Lachlan River at Cowra, northeast of Young, advised new farmers on the western slopes of the need to remove the 'suckers' and thereby stop ringbarked trees regrowing. Axes shaved away emerging branches and leaves. To ensure the death of the trees, farmers sometimes had to undertake 'sucker-bashing' three times, Sutton explained.[66]

Occasionally there was no need to return and remove suckers. According to one practiced land clearer, wet and stormy summers were ideal for ringbarking: 'Followed by a crisp cold winter, there seems to be less chance of regrowth in any form'.[67]

Decades before William Farrer became a famous wheat breeder early in the twentieth century, the scientist explained the advantages of ringbarking for sheep farming.[68] Ringbarking, he proposed, allowed graziers to boost sheep numbers fourfold. Sheep grew more wool on ringbarked country,

> for, not only will ring-barking have allowed more grass to grow, but as the sun will now be able to get at it freely, it will be made

nourishing, and will contain more of the wool-forming nutriment; and where, in all likelihood, a less quantity of poor sour grass grew, there will now be a larger supply of good, sweet, nourishing pasturage.

Farrer cautioned against excessive ringbarking. Trees 'here and there' gave shelter to stock, and attracted rain. He suggested graziers save

> for ornament any well-shaped trees, or any trees that are valuable for their timber – to leave the country, in fact, wooded like an English park.

Sutton agreed with Farrer. Ground 'sweetened' as leaves and bark dropping from dying trees bolstered the mineral and organic content of soil, and as sunlight swept the ground. Sutton suggested that farmers graze ringbarked country as they waited for stricken trees to die. Several years on, dead trees stood in soil ready for crops.[69]

At Mimosa, southwest of Temora, a local blacksmith invented the 'Forest Devil', a mechanical device for removing stumps and pulling down ringbarked trees.[70] One winter evening in 1884, after a day of ploughing matches, several hundred people witnessed the trial operation of a Forest Devil on a farm near Cootamundra. Men fastened long chains to a dead tree and adjusted a lever. Within eight minutes the tree was down. 'These machines are a wonderful improvement', the *Cootamundra Herald* exclaimed, 'and no farmer who has much land to clear should be without one.'[71] Using a Forest Devil, three workers could remove thirty trees a day. Farmers readied cleared paddocks for ploughing by 'grubbing' out stumps and roots. To save time and effort, some farmers instead left the roots in the ground and used a 'stump-jump' plough. The 'more careful farmers', Sutton observed, did not advocate this faster, cheaper method of clearing and farming.

Ringbarked forests turned grey and lifeless across the western slopes of New South Wales as agricultural settlement intensified late in the nineteenth century. Farms and towns emerged beside stands of dead trees and remnant woodland. Mary Gilmore recalled one paddock of 'Riverina red earth which cracked and crumbled, bone-dry in the heat', where the hole of a trapdoor spider descended between native grass tussocks. A wheat crop grew nearby, and bushland remained beside the dead trees:

> Near where the trap-door spider dwelt, the Willy Wag-tail used to swing, like a swaying leaf, on a clod in the long dry furrow of an adjacent wheat paddock which ran down to a shallow creek where water flowed only when rain was heaviest. In spring the cockatoos watched that field in its sprouting green, massed in clusters on dry ring-barked trees which had not yet been grubbed, and, at sight or sound of a gun, flying to the shelter of the bush behind the fences.[72]

West of Temora, retired farmer Ian Thompson worked land that was once part of Mimosa station. He told me a story about his grandmother, Caroline MacLennan, who regularly played the organ at the local church, a small weatherboard building among paddocks and ringbarked trees. One evening, MacLennan walked away from her farmhouse towards the church, about a mile away. As darkness fell, she lost her way inside a stand of ringbarked eucalypts. Neighbours travelling to the service spotted her lantern moving amid the dead wood. They found her confused, disoriented by the uniform tangle of grey trunks and broken limbs.

Escorted to the church, Caroline MacLennan regained her composure before taking her place behind the church organ.

~ ~ ~

A groomed paddock separates a forest remnant from the busy Olympic Way southwest of Bethungra. Ringbarkers swung axes here generations ago. Stumps and fallen timber lie beside weathered trunks persisting upright. Some elderly box trees remain, suggesting the ringbarking team never returned to cleave away suckers emerging below axe wounds. Open paddocks stretch away from the few hectares of rotting wood and old trees.

Ian Thompson told me that ringbarked forest remnants were common once. Draught horses grazed native grasses and forbs between dead trees and fallen timber. Ian was sorry to see the forest remnants vanish as farming became more intensive and mechanised after World War II. Paddocks of dead trees, fallen timber and grass tussocks had given refuge to small marsupials and ground-dwelling birds.

The stand of old box trees near the Olympic Way once joined a forest cloaking the Bethungra squatting run. Frank Cowley told a visiting journalist that his station was 'a complete wilderness' when he purchased the property in 1876. Cowley had amassed money working as a contract surveyor on the Great Southern Railway extension south through Bethungra towards Albury. He decided to become a grazier. Despite a procession of dry years since he took up Bethungra station, the journalist noted, Cowley

> worked steadily onward – fencing, building, conserving water, ringbarking and clearing – undeterred by all sorts of difficulties, until now his estate is a pleasure to behold.

Governments encouraged land clearing. Laws passed in 1881 deemed ringbarking an 'improvement', allowing pastoralists to claim compensation from farmers who selected ringbarked land.[73]

Cowley built a handsome red brick homestead for his family on a sharp rise near the railway line, and renamed the run Bethungra Park. According to the *Sydney Mail*, people thought Cowley's property 'one of the most highly improved' in southern New South Wales.[74]

Early in the twentieth century, Bethungra Park appeared in *The pastoral homes of Australia*, an expensive publication featuring large and prosperous grazing stations.[75] Charles Westmacott bought Bethungra Park in 1898, the volume records. Once forested, the entire station was now ringbarked. More than two thousand acres (roughly eight hundred hectares) lay clear of dead trees, ready for the plough. 'The boundaries are wire-netted, and the run is subdivided into five blocks, also securely netted', readers learned. One photo shows four draught horses dragging iron harrows across a bare paddock beside a stand of ringbarked trees. Dust rises behind. Other photos show merino sheep, iron sheds and white box trees. In several images, steep ranges dense with eucalypts curve across the skyline.

The pastoral homes of Australia paid homage to 'those courageous men who faced the bush when it was as the first white men found it.' These 'venerable pioneers' had 'actually seen, or have heard their forebears tell of the days before railways pierced the bush.' They were men who 'got their stock together and penetrated a lonely land'. In the early years there was 'no noise of warfare' across the inland, though 'a conquest was none the less achieved', as men steadily triumphed 'against nature's silent forces'. Aborigines 'caused much suffering and trouble' for a time, 'but they quickly faded away', and their demands and actions were easily forgotten. Barriers to progress did not daunt the pioneers. Hard work and ingenuity gave rewards:

> Lack of water was one of the early obstacles to be overcome and this volume bears witness to the success achieved. Then there was the bush, the heavy gum trees which would not permit grass to grow underneath them. But they were in time replaced by grass, not by hewing down each tree as is necessary in nearly every other country but by simply nicking out a circle of bark from around each trunk. When the timber died, grass and nutritious herbage at once appeared and the carrying capacity of the country was increased immeasurably.

'Clearing has been a major factor in the progress of Australian agriculture', eminent agricultural scientist Colin Donald observed in 1982.[76] Donald described dramatic changes in land clearing methods after World War II. Landholders abandoned labour intensive practices like ringbarking. Instead, they used bulldozers, bulky chains drawn between tractors, and new industrial poisons. Mechanised farming systems developed by agricultural scientists and technologists in league with farmers drew immediate and substantial income from land cleared of grassy woodland. Scientists advised farmers to sow exotic pasture species and apply superphosphate fertiliser. Crops and pastures grew vigorously as soil nitrogen and phosphorus levels soared.

From the deep verandahs and iron lace of Bethungra Park homestead, a steep range of blue hills defines the eastern horizon. A gap appeared on the skyline late in the 1970s. In a steep back paddock, clearing contractors walked through a shady forest of drooping she-oak, black cypress pine, red box and stringybark. They cut into the bark of trees and injected systemic herbicide. Later, planes dropped clover seed and superphosphate over the stony range of dying trees. When suckers rose from tree roots and trunks, the contractors returned and finished the job. A handful of shallow-rooted, annual pasture species replaced a diverse matrix of native annual plants and deep-rooted perennial grasses, shrubs and trees.

Without a plant community adapted to conserve moisture, the range of hills no longer absorbed and held rainfall. Water shifted down and through stony hillsides. On the slopes below, farmer Owen Whitaker watched scalded areas spread as salty groundwater surfaced. After heavy rain, sheets of water gathered across the steep hills – now bare of grass tussocks, shrubs and fallen limbs – and tore creek banks away. A bushfire erased the grey, broken remains of the forest a decade after the poisoning, Owen told me. In winter and spring, sheep graze a fertilised pasture of exotic annual grasses and weeds. In summer, the bare hills bake.

Transcendence and war

Abundant rain torments farmers beside Bland Creek in the Morangarell and Grogan districts, north of Stockinbingal. Water takes months to evaporate and drain away from heavy clay soils and flat paddocks. Crop yields are often lower than in adjacent areas, where soils drain more quickly. According to local farmers, compaction by machinery and deteriorating soil structures worsens the situation. They rush to sow crops when autumn rains begin. Paddocks soon turn too boggy for heavy farm machinery.

In 1998, a particularly wet year, waterlogging destroyed many crops beside the Bland. Agricultural scientists began working with local farmers to solve the problem. According to the Grains Research and Development Corporation, there was potential 'for large productivity improvements' along the Bland.[77] Funded by the corporation, the scientists developed strategies to help farmers overcome the limitations presented by the land. Deep ripping and the application of gypsum, they discovered, broke soils apart and improved drainage. To soak up excess water, they suggested landholders plant lucerne, a perennial pasture with deep roots.[78]

Bland Creek rises in the Dudauman and Bauloora ranges, northwest of Cootamundra, and flows past the village of Stockinbingal. The major waterway curves through paddocks of dark, swampy earth towards Lake Cowal, south of the Lachlan River.

European colonists found the fertile, watery area teeming with diverse life. Mary Gilmore's father told her the Wiradjuri name for Lake Cowal 'could be translated almost as "the garden of Eden" it was such a place of singing birds and flowers.' She remembered 'the flowers like a carpet, the wading and swimming birds in thousands, the bittern booming in the night.'[79]

In the middle of the nineteenth century, Alexander Mackay managed Stockinbingal run on Bland Creek upstream from Lake Cowal. He built a sturdy dam wall across the Bland. Water banked up for two miles. 'I know that, when I was a boy', wrote Alexander's son Kenneth, 'it was full of fresh-water cod, and that the blacks used to catch them by mudding the water.'[80] According to local historian Ethel West, the area just west of Bland Creek at Stockinbingal 'was covered by shallow water and reeds, and this marsh was the home for wildfowl of all kinds.' She thought the suffix '–bingal' a Wiradjuri term for 'marsh'.[81] Wet conditions made the area 'almost impassable' when James Larmer journeyed north along the Bland from Stockinbingal in the winter of 1848. The surveyor carefully recorded local placenames from Wiradjuri people. One major lagoon beside Bland Creek was called Moonbooga.[82] In 1879, a travelling reporter described the abundance of water along the Bland:

> Usually it is nothing else but a chain of waterholes, but these are often large and deep, retaining an abundant supply of water in all seasons. As it drains a large area of plain country, it is subject in wet seasons to high and dangerous floods, to the great inconvenience of that portion of the inhabitants who have made their homes in too close proximity to its banks.[83]

Relations of Mary Gilmore once held Morangarell station on the Bland downstream from Stockinbingal. According to Gilmore, 'Morangorell' meant 'the nesting place of the waterfowl'.

In the 1860s and 1870s, before a tide of agricultural development swept the southwest slopes, Wiradjuri gathered regularly at Morangarell to catch waterbirds and hold ceremonies.[84] Sara Hawkins, born in 1862 on Curraburrama station, downstream from Morangarell, recalled those 'days of prolific seasons' in her youth, when 'the waving grass on some places on the Bland would completely hide a horseman'. The country of her childhood was wet and lively:

> Emus and Kangaroos were numerous and there were plenty of wild fowl and fish. Blacks would come to the homestead with fish strung on green rushes and ask for food and tobacco in exchange. They were rather numerous then, and wore possum rugs and blankets pinned round them with wooden pins. They carried their war weapons and always had a number of dogs.[85]

When I visited Morangarell recently, a farmer described to me the Bland Creek swamplands he knew as a child. Among 'bulrushes' lived swans, black duck, grey teal, migratory snipe and pelicans. In the 1950s and 1960s, his father and other farmers built drains through swampy paddocks with new earthmoving equipment. They grew crops on the drained land. Birds, frogs and rushes vanished. 'Now I'm not an environmentalist', the farmer told me, 'but I don't think they should've drained those swamps.'

Today, strategies offered by agricultural scientists help Bland Creek farmers further transcend the particular nature of the watery area.

Since the emergence of agricultural science as a distinct field of inquiry late in the nineteenth century, the discipline has been defined by a belief that humans can defy natural limits and boost primary production through the simplification and strident transformation of local ecologies. In 1964 Colin Donald, then Professor of Agriculture at the University of Adelaide, explained that there were two ways humans might build 'more rewarding' relations with their environment.[86] Man 'may seek to adapt himself more closely to the conditions he finds', wrote Donald, 'or he may attempt to modify the environment to his particular needs.' For thousands of years, farmers had substantially modified land for cropping. Now with the help of modern science graziers too could 'transform the environment and productivity' of regions, Donald observed.

Agricultural scientists rejected the idea of adapting to local conditions and cast the natural characteristics of rural Australia as problematic barriers to production. 'The history of Australian agriculture', agricultural economist Bruce Davidson wrote in 1985, 'is largely a story of developing new technologies to overcome problems as they arise.' According to Davidson, limits to development included those 'caused by the physical environment', by the natural realities of the land. He considered even 'the location of the continent' an obstacle to successful agriculture.[87]

Agricultural scientists viewed disorders induced by earlier scientific farming techniques as new challenges requiring clever solutions. In 1955 Robert Watt, Emeritus Professor of Agriculture at the University of Sydney, honoured those plant breeders who worked with 'tenacity and persistence' to deliver improved varieties and bulkier harvests 'in spite of soil erosion and obvious decrease in soil fertility in many regions'.[88]

Few agricultural scientists, it appears, questioned the wisdom of strategies to erase and transcend the natural realities of rural places. Some did admit to a sense of foreboding. Agricultural scientist Robert Noble joined the New South Wales Department of Agriculture in 1913, becoming director of the department almost three decades later.[89] During his career, Noble witnessed profound changes in Australian farming systems. He gave the annual Farrer Memorial Oration at Hawkesbury Agricultural College in western Sydney when he retired in 1959.[90] He reflected on the great faith people had in the ability of interventionist science and technologies to deliver an abundance of primary produce. 'There are some who regard it as heresy for anyone to suggest that there are any limitations to Australia's production potential', he told the audience. 'Mistakes were made' as settlers developed agricultural regions,

> and it is easy for us for example to assess now the losses – in some cases the irreparable losses – which have occurred through erosion of our soil, the most precious of our natural resources. Let us hope that we too, are not, quite unknowingly, making mistakes for which future generations may have to suffer.

The New South Wales Department of Agriculture employed some of Australia's first agricultural scientists. Formed in 1890, the department hoped

> to help those now on the soil, to educate their sons and daughters who will succeed them, and to offer every facility and encouragement to wider and more intelligent occupation of the still unsettled tracts of this great colony.[91]

Two years later, the department opened an experimental farm near Wagga, north of the Murrumbidgee River beside Houlaghans Creek. Workers ringbarked, grubbed and burned a forest of red gum, white box and black cypress pine. Here, young farmers studied alongside scientists. The students learned agricultural theory and practice. Scientists trialled varieties of wheat, grapes and other crops thought suitable for inland districts.[92]

To further promote agricultural science and learning, the New South Wales government opened other experimental and educational institutions on the southwest slopes and plains. After widespread lobbying by local farmers, the Temora Demonstration Farm opened in 1912

> to examine the agricultural problems of the district and to demonstrate to farmers the best ways to manage crops which were suited to the district.[93]

Agricultural scientists at Temora tested varieties of wheat and oats and explored new methods of crop rotation and pasture improvement. In 1934, investigations began into the establishment of subterranean clover pastures nourished with superphosphate fertiliser, a method pioneered by research scientists in the western districts of Victoria.[94] In the 1930s and 1940s, 'sub and super' transformed farming systems across southern Australia. Alex Baldry, a retired Wallendbeen farmer, told me of his father's excitement after following Department of Agriculture advice to sow sub-clover and spread superphosphate – the clover grew so tall. Fertilised clover supported many more sheep than the brome, barley and kangaroo grasses it replaced. Phosphate fertiliser fuelled clover growth as autumn and winter rains fell. A leafy annual, sub-clover set seed during spring then died away. Over summer, sheep grazed dry clover herbage and nutritious deposits of clover seed.

Subterranean clover is a leguminous plant. Bacteria form nodules on roots and fix nitrogen captured from the air into the soil. After World War II, rising soil fertility rates and developments in agricultural machinery and chemicals encouraged farmers to commit more paddocks to cereal crops. The area sown to wheat on the western slopes of New South Wales rose almost fourfold between 1945 and 1980.[95]

Aesthetically, farmland and grazing paddocks changed profoundly. 'The country looked much wilder years back, before the war', Alex Baldry said. He remembered some native grasses growing tall, others short. Walking through paddocks as a child, he often found bird's nests hidden amongst grass tussocks. Weeds were a problem. Chinese thistle, a plant with spiked yellow blossoms, was particularly hard to control. In later decades, Chinese thistle and other problem plants vanished as farmers took advantage of new herbicides. Alex led me down an interior hallway to see an old photograph in a timber frame, hanging from the picture rail. In the picture, his homestead and garden sat on a hill ruffled with native grasses.

In the twentieth century, a complex dynamic of economics and culture drove settlers to transform rural places across the western slopes. Perceptions of indigenous plants and local ecologies as deficient and problematic enabled change. Australian rural terrain, agricultural scientists emphasised, was characterised by 'recurrent droughts', 'sparse and erratic rainfall', soils of 'extreme poverty', 'useless scrub and poor forest', 'inferior natural swards' of perennial grasses 'unadapted to grazing by sheep and cattle'.[96]

Agricultural scientists saw farmland as a malleable resource available for national and industrial purposes. 'No one argued that we should accept this poor, old continent for what it was', remembers retired agricultural scientist David Smith. 'It was ours to improve, to manage. We were to take what was and use our knowledge to make gain for our nation and humanity'.[97] Devaluation of native species, of specially adapted and integrated components of living systems, of plants and animals central to Aboriginal culture, allowed dramatic transformation. 'Don't be hung up about natives', Smith told a Murrumbidgee Landcare Association gathering in Cootamundra recently. 'A native is simply a plant that was here before 1788.'[98]

Representations of land as inhospitable undermined motivation and potential for relationships of intimacy and care. 'AS TOUGH AS THE COUNTRY IT HAS TO FENCE', an advertisement for A.R.C. Weldmesh rural products declared in 1985.[99] Casting places as 'indifferent or malign' served colonial purposes, suggests literary scholar David Tacey.[100] Such oppositional representations justified and promoted efforts to harness rural places for high-output, export-oriented production.

In 1967, agricultural scientist Eric Underwood described the 'environmental improvement' science delivered to the harsh terrain of rural Australia, 'a record of which any country could justly be proud.'[101] Underwood drew upon and reinforced nationalist myths of heroic colonists doing battle with the land. He implicitly denied the experience and ecological understandings of Aborigines, people who saw the same terrain as familiar and nourishing:

> The pioneer settlers of this country found themselves in an environment which was always strange and often harsh. There was no body of local practical experience upon which to draw and no research and extension services upon which to lean. Such challenging circumstances called for courage, initiative and endurance of a high order, as well as a capacity for hard physical work. These qualities were abundantly present in the minds and bodies of the early farmers and graziers and their wives. Such qualities are still required and still exist in the men and women on the land in many parts of this continent.[102]

Agricultural science offered settlers power over natural forces they considered troublesome.

Promises of mastery defined the philosophical framework of Enlightenment science that emerged in western Europe in the seventeenth and eighteenth centuries. René Descartes, a leading progenitor of Enlightenment rationalism, drew a sharp distinction between physical matter and the human imagination. People could apply objective scientific knowledge, Descartes argued, understandings abstracted from the distorting particularities of time and place, to achieve 'mastery and possession of nature'.[103]

The Enlightenment basis of agricultural science is reflected in the academic cultures of Australian agricultural schools and colleges. Yanco Agricultural High School opened northwest of

Narrandera in 1922. The Department of Education appointed Ernest Breakwell as principal.[104] Formerly an economic botanist in the Department of Agriculture, Breakwell sought to attract boys from urban and rural backgrounds. A feature on Yanco Agricultural High School appeared in a Sydney newspaper a year after it opened. No young man, readers were told,

> whether he be a town-bred boy or a farmer's son, should be denied the opportunity of obtaining all possible information and scientific training in the greatest of all industries.

Keen competition for primary products made thorough training in agriculture a necessity for boys with an interest in farming. 'To no one is there more significance in the old school aphorism "knowledge is power" than to the young man who is to be a farmer', the newspaper declared.[105]

Enlightenment scientists believed mechanistic laws governed natural systems. Nature was knowable and controllable, not complex and dynamic. Agricultural science became 'a world force' in the first half of the twentieth century, observed Otto Frankel, chief of the Division of Plant Industry at the Commonwealth Scientific and Industrial Research Organisation (CSIRO) from 1951 to 1962, helping 'to conquer the earth for man's use.'[106]

Reflecting on ecological alterations imposed across rural southern Australia since World War II, David Smith celebrated a history of 'conquest' achieved by 'visionaries' and 'champions'.[107] Smith praised fellow agricultural scientists, men like Eric Underwood. In 1928, Underwood had left the University of Western Australia with a first class honours degree in agricultural science and travelled to England. At the University of Cambridge, the young scientist had studied factors influencing milk availability in ewes and the nutritional effects of nitrogen fertilisers.[108] Back in Australia, Underwood had become an expert on the nutrition of grazing animals. In 1976, the Royal Agricultural Society of England elected him an honorary member.

The National Library of Australia holds Underwood's papers. One photo in the collection shows a distinguished figure glancing at notes through dark-rimmed glasses, grasping a lectern and speaking into a microphone. Perhaps the photograph was taken at the University of New England in 1967, when Underwood argued for the maintenance of close ties between farmers and agricultural scientists:

> Producer and scientist will need to maintain the confidence in each other that has been such a significant feature of the past and the community as a whole must be kept informed of the rich dividends which have come and which can come from investment in science and its application to our primary industries. It is this investment which has contributed so much to the great tradition of the Australian people – the mastery of the land.[109]

Agricultural scientists and technologists developed radical new systems of primary production in the second half of the twentieth century. Chemical farming methods brought an end to the practice of repeated cultivation to control weeds and prepare paddocks for sowing, a practice that exposed farmland to significant erosion. Stubble retention and minimum tillage, made possible by broad-spectrum herbicides and more powerful tractors, returned organic matter and structure to soils. On the southwest slopes, the frequency of severe erosion events declined.

Successful conservation and restoration of farming soils did not, however, reflect changes in the dominant imaginative framework of modern agriculture. Economic factors ensured a forceful quest for mastery over natural systems still characterised Australian farming.

After the end of World War II, declining terms of trade caused farm input costs to rise persistently faster than prices received for produce, a phenomenon called the cost-price squeeze. Farmers with enough skill, education and financial capital made adjustments. Many bought or leased properties from neighbours unable or unwilling to hang on. Each year since the middle of the 1950s, about two thousand Australian farms have disappeared into other holdings.[110]

In the 1970s and 1980s, as Australian governments withdrew a range of economic devices protecting rural industries, landholders could not easily explore alternative ways of engaging with farmland. In real terms, total farm debt in Australia rose from almost ten billion dollars in the late 1970s to more than twenty billion two decades years later.[111] Unsurprisingly, ecological problems like soil acidification and dryland salinisation emerged as major threats to farmland productivity over the same period. Across expanding acreages, land managers intensified farming practices. Seeking profits, they applied new mechanical and chemical technologies to boost yields. Natural systems began to falter.

An 1878 edition of *The Garden and the Field*, a South Australian agricultural journal, describes a process of 'warfare' waged against rural places by agricultural societies, local organisations promoting scientific farming methods:

> We look upon an Agricultural Society in the light of an army of soldiers engaged in an effort to conquer and 'civilise' the land – that is, to subdue it and render it subservient to the uses of man.[112]

People increasingly applied the language of war to describe Australian agriculture as economic pressures and industrial production systems intensified in the second half of the twentieth century. Crop monocultures are particularly vulnerable to weed competition, insects and disease. Powerful means of protection are necessary, furthering the use of war rhetoric. 'Don't be in doubt as to how to wage war against profit devouring weeds and insects', farm chemical company Wilcox Mofflin told farmers in 1955. 'Simply describe the problem in a letter to Wilcox Mofflin and let our Weed Experts supply the effective answer.'[113]

Force replaced mindful attentiveness to the land. 'The skies above the NSW cropping belt are buzzing this week as squadrons of agricultural aircraft are hurled into a massive pest and disease control offensive', *The Land* newspaper reported in October 2004 as stripe rust, aphids, heliothis bugs and mice infested grain crops struggling through a dry season.[114] The brand names of farm chemicals used today reflect the forceful approach of modern agricultural systems: Advance, Ally, Arsenal, Ballistic, Barrel, Baton, Broadside, Broadstrike, Clearfield Advantage, Detonate, Flame, Fusilade, OnDuty, Rifle, Scud, Showdown, Sniper, Stomp, Verdict, Wipeout.

As competition strengthens in the global marketplace, manufacturers offer rural landholders a tighter grip over natural systems. Advertisements for farm chemicals and machinery promise farmers 'more power' and 'more control'. Today, single tractors with the pulling power of five hundred horses are on the market. 'Unleash extreme power and productivity' invites the manufacturer of Powerful Knights, a new line of tractors, in a recent newspaper advertisement. 'Like legendary knights of old', reads the advertisement beside an image of giant tractors and a bare paddock, 'these all-new powerfully built tractors command respect.' A turbocharged engine with '450 horsepower waiting to be unleashed' propels the largest machine available. The new tractors are weapons of war, and natural terrain is cast as hostile: 'With record-breaking power, torque and strength these formidable machines can do battle in the harshest conditions'.

Battling the land: 200 years of rural Australia, published in 1999, seeks the 'grand themes' of Australian rural history and culture. Rob Linn, author of the celebratory book, describes 'a prolonged battle between people and the land'. *Battling the land* reveals and reinforces heroic narratives of rural colonisation. Farmers, struggling for the national good with variable and brutal natural forces, are granted moral status, undermining necessary critique. Conflict between settlers and nature, Linn observes, lies 'at the heart of any true understanding of the story of rural Australia'. Today, as ecological disorders threaten the productivity of farmland, Linn sees modern science and the Landcare movement as offering 'the means of overcoming adversity'. Linn presents ecological sickness as another barrier to production, another challenge facing scientists and farmers.

Even within Landcare, a movement often represented as symbolic of deep attitudinal change, the established project of mastery appears to prevail. In the context of modern agriculture, notions of settlers 'battling' rural places underlie the use of terms like 'sustainability' and 'natural resource management'. Farmland, bound to the global marketplace, is considered primarily a 'natural resource' available for commercial activity. The rhetoric of 'natural resource management' casts people outside nature, above and in control of land and other species. Synonyms of 'manage' listed in *The Oxford thesaurus* include 'administer', 'boss', 'control', 'direct', 'discipline', 'govern', 'handle', 'head', 'lead', 'manipulate', 'mastermind', 'oversee', 'preside over', 'regulate', 'rule', 'run', 'supervise', 'take care of' and 'watch over'.

Attitudes dominant in natural resource management and modern farming block dialogue with rural places. Nature is often seen as a troublesome opponent. Profitable production, it is assumed, requires force and monological relations of power. Unfortunately, the ecological disordering of farmland has not evoked widespread critique of the complex economic and cultural dynamics driving quests for mastery over rural places and the natural forces of the land.

~ ~ ~

The small town of Henty grew beside Buckargingah Creek in the late 1870s, as the Great Southern Railway extended south from Wagga. To generate steam, railway engines need regular supplies of water. Workers and horse teams excavated a major reservoir on Buckargingah Creek. Pipes drew water into an overhead tank. A small town grew as selectors carved farms from pastoral runs.

Today, the streetscape of Henty reflects the commercial dynamics behind the transformation of local hillsides and creek flats into farmland. Concrete silos form a backdrop to solid bank buildings.

In the main street, a sign directs visitors to a town park to see the Headlie Taylor Header Memorial. An old mechanical harvester, carefully restored to working order by members of the United Farmers and Woolgrowers Association, stands on display inside a small shed with transparent walls. Early in the twentieth century, local farmer Headlie Taylor developed a mechanical device to harvest and process wheat crops flattened by heavy storms. His invention attracted great interest. In 1916, he joined the Sunshine Harvester Works in Victoria, a prosperous manufacturing firm owned by HV McKay, famous inventor of the Sunshine stripper harvester. In the Sunshine factory, Taylor further developed his popular machine. 'The Taylor Header is regarded as the greatest single contributing factor to the development of the world cereal industry', a sign reads above the restored Headlie Taylor Header.

The association of the Henty district with farm machinery extends beyond the Headlie Taylor Header. Farmers on the south-west slopes first attended the Henty Machinery Field Days at the local showground in 1963. They listened carefully as machinery companies touted new products, inventions to boost the efficiency and profitability of farming systems and counter the effects of declining terms of trade, a worrying and persistent trend. In 1977, organisers secured title over a travelling stock reserve east of Henty beside Buckargingah Creek. The reserve 'was totally filled with timber', a local historian notes,

> and as such, developed from that time, has seen the extraction, burning and preparation of the site to where it is now, one of the most attractive Field Day venues in Australia.[115]

As the twentieth century closed, intensifying competition in the global marketplace and a tightening cost-price squeeze ensured growing popularity for the annual event. 'Independent research shows that farmers are hungry for technical information about new technology that can improve their productivity and reduce costs', a journalist observed in 2001 as thousands of people gathered at the Field Days site.[116]

Wheat crops were dark green and canola was starting to flower when I drove to Henty Field Days two years later. Across three September days more than sixty thousand people attended. Exhibitors displayed 250 million dollars worth of machinery.[117] Reflecting established traditions of agricultural science and technology, organisers promised farmers transcendence over the natural character of Australian farmland:

> Henty is a massive outdoor supermarket for leading edge farm machinery, products and services from North America and Europe and showcases Australia's unique engineered and technical solutions to the challenges of farming on one of the world's driest continents.[118]

Frogs called from Buckargingah Creek as I entered the Field Days site. Scattered yellow and grey box trees rose above pavilions, machines and the mass of visitors. One marquee housed a photographic exhibition of old farming scenes. I examined a picture of horse teams pulling reaping and binding machines through a ripe wheat crop. Workers gathered sheaves and placed them upright, building a series of stooks. Another photo showed six men loading hessian bags filled with grain onto the back of a truck. In spring sunshine outside the marquee, farmers inspected new machines with names like Maxxima, Challenge, Speed Drill, Agmaster, Magnum, Dominator – names redolent of economic demands for immediate, high-output production, of pressures sustaining a desperate quest for mastery over rural terrain.

Australian farmers now carry a combined debt level of almost thirty billion dollars.[119] On the day I visited Henty Machinery Field Days, agribusiness bankers inside pavilions sold farmers 'innovative and customised financial solutions'.

I joined a crowd to hear Ian Macdonald, New South Wales Minister for Agriculture and Fisheries, open the popular event. He spoke of an urgent need for more applied research and new technologies to keep farm productivity rising.

Each spring at Henty, the monumental granite rise of Cookarbine Hill forms a backdrop to flapping banners and rows of grain

River red gums, Dudal Swamp, Henty district.

augers angled skyward. Buckargingah Creek gathers water from the steep slopes of Cookarbine Hill and Buckargingah Sugarloaf. The waterway meanders westward through paddocks, past the Field Days site and the town of Henty. Beyond the settlement, Buckargingah Creek flows into Dudal Swamp, a wide depression studded with old river red gums.

Generations ago, locals noticed the disappearance of hillside springs after ringbarking teams denuded Cookarbine Hill. Instead of soaking down through porous granite and seeping from slopes, rainfall drained swiftly across hillsides bared of trees, shrubs and fallen limbs, of the textured surface that formerly slowed and captured water. Early in the twentieth century, when Dudal Swamp flooded during particularly wet seasons, Henty residents gathered for boating events. As decades passed, sandy material eroding from hillsides and paddocks spread across the swamp. Red gum saplings grew, preventing even the smallest boats from sailing.[120]

Late in the afternoon, I drove a short distance along a sandy track south of Henty, alongside Dudal Swamp, and started walking. Cattle occupied green swampland paddocks. Black and grey calves ran about, tails lifted in play. The air smelled moist. I passed river gums sinuous and tall, solid forms holding centuries. A steady hum of trucks and cars carried from the highway. White-winged choughs made raucous sounds in grey box trees nearby.

On the track I noticed a pale stone, and stopped to pick it up. With a ground edge, the small tool looked like a miniature greenstone axe. I held the worked stone in the palm of my hand, so unlike the bulky, complicated instruments on display at the Field Days.

As I gazed at the stone beside the ancient red gums of Dudal Swamp, it seemed to suggest other ways of seeing and engaging with land. While modern industrial devices are mechanically elaborate, they crudely impose dramatic change. Understandings of nature as knowable and controllable continue to drive the development and application of many agricultural technologies. Unresponsive to the shifting, varied needs of local ecologies, modern machinery requires and imposes a simplification and fragmentation of natural patterns.

Wiradjuri people approached the land differently. Subtle responses based on intimate understandings of the dynamic and expressive life of places fostered a diverse abundance of swamp and grassy woodland species. Biological diversity and ecological connectivity made land resilient and maintained natural productivity and stability.

Powerful machines developed and applied by British settlers emerged from different cultural traditions and histories. Narratives of technological and social progress, perceptions of a fundamental divide between humans and nature, commitments to global ties of trade and culture, and mechanistic understandings of natural processes led settlers to transform rural places in strident ways. Dryland salinisation, soil acidification, premature deaths of paddock trees, declining woodland bird populations, waterway pollution and other ecological problems present today in Australian agricultural regions are unfortunate, unforeseen consequences of major interventions in dynamic and intricate systems of nature.

elsewhere

Pathways

In her rich and insightful *Old days: old ways – a book of recollections*, Mary Gilmore describes a conversation she overheard as a child, standing by her father. Donald Cameron and a senior Wiradjuri man were talking about the different types of ships known to land in northern Australia.[1] When boats with brown and single mat-like sails approached the shore, Aboriginal clans welcomed the sailors from Macassar, an island on the other side of the Arafura Sea. But when a ship appeared with tall masts and many white sails, Gilmore remembered the Wiradjuri elder saying, women and children went inland. Europeans, unlike Macassans, brought disease and violent death.

In the Murrumbidgee region, so far south of the warm seas of northern Australia, Wiradjuri knew in detail the differences between European and Macassan ships. There are Wiradjuri names for Torres Strait and Cape York.[2] Across Australia, people carried stories and ideas with ochres and greenstone axes, along pathways of ceremony, communication and trade.

British colonisation bound the southwest slopes of New South Wales to faraway cities and export destinations, to global networks of trade and power. Routes of commerce and migration enabled dramatic social and ecological change. Squatters made tracks northeast across the southern tablelands to deliver fat cattle and wool bales to Sydney markets.

Unfamiliarity with land made settlers vulnerable as they traversed the western slopes in the early decades of settlement. John White came to the Young district in 1828 to help his brother James found Burrangong squatting run. A century later, his daughter Sarah Musgrave published *The wayback*, an account of life at Burrangong when colonisation of the eastern Riverina began.

Six years after he arrived, Musgrave explained, her father became lost in dry, scrubby terrain. A boy rounding up cattle encountered John White's body nine days later, limp and torn by dingoes. Sarah Musgrave told stories of other men vanishing into bushland. A shepherd lost his bearings east of Burrangong homestead in the area where John White perished. To stay alive, he cut pieces from the tails of his two sheep dogs and drank their blood. He stumbled into the homestead six days later. According to Sarah Musgrave the faithful dogs, thirsty and maimed, stuck by the man throughout.

After the death of his brother John, Burrangong squatter James White and Wiradjuri elder Cobborn Jackie marked the eastward track. They cut blazes into tree trunks to signpost the route through the bush. Cobborn Jackie was a skilled 'bush surveyor', wrote Musgrave, 'wonderful in the art of straight lines.' He helped White survey many tracks. Pathways determined by White and Cobborn Jackie joined routes defined by other Wiradjuri for neighbouring squatters. Rough tracks traversing and linking squatting runs eventually became main roads, Musgrave noted, bitumen thoroughfares today busy with cars and trucks.[3]

In the early decades of colonisation on the southwest slopes, squatters relied on bullock wagons to transfer goods to and from urban markets and seaports. As Melbourne grew in the 1830s, some Riverina squatters made the journey to the southern port settlement instead of Sydney. Drovers delivered cattle and sheep fattened on kangaroo and wallaby grasses to city saleyards and slaughterhouses. Wagons carted away heavy bales of wool and returned to stations laden with supplies. Goods ordered in 1850 from Sydney for Wallendbeen station included woolpacks, sheep shears, tobacco, children's boots, rice, cooking utensils, eighteen dresses of various print, vegetable seeds, six shepherd's coats, tools, blankets, an account book, saddles, a gun and lead shot, and one table cloth.[4]

Rough roads meant long journeys. A return trip by bullock wagon to Melbourne or Sydney could take months. Riverina colonist James Gormly remembered a treacherous route alongside the Murrumbidgee between Wagga and Narrandera. In wet seasons, travellers encountered billabongs full of water. Seven people drowned along the road one winter in the 1850s.[5]

Gormly described one journey to Melbourne from Muttama station, between Cootamundra and Gundagai, by squatter Francis Taaffe and several stockmen in 1838 with a mob of cattle.[6] On the way, the men learned of the slaughter by Aborigines of at least eight stockmen beside the Broken River southwest of Albury, where pastoralists were attempting to secure land from local clans. After the cattle sale, safe in Melbourne, Taaffe's workers refused to make the dangerous journey home. The squatter departed on Barebones, his thoroughbred horse. As settlers abandoned stations throughout the Broken River region, Taaffe avoided the main road and spurred his horse through the bush. He urged Barebones to swiftly swim rivers. Every night the squatter walked to rest his weary horse. The pair reached Muttama station in just four days. Overwhelmed with gratitude, Taaffe never saddled Barebones again. Years later, Gormly saw the old horse grazing contently in a paddock beside Muttama homestead.[7]

Steamboats began plying the Murrumbidgee River in the 1860s to service towns and stations. Boats entered the river at its confluence with the Murray, near Balranald, and journeyed upstream to Wagga. Steamers were faster and cheaper than bullock wagons, and more comfortable than coach travel. As Narrandera historian Bill Gammage explains, nothing else could permit travellers a bath or a drink, or let them laze back watching the scenery without being jolted savagely about, and nothing else could link the bush so quickly and cheaply with the world.[8]

Boats and barges carried produce loaded from Murrumbidgee wharves thousands of kilometres west to Goolwa, at the mouth of the Murray River in South Australia. A horse-drawn railway linked Goolwa to Port Elliot, where workers stacked wool bales onto ships bound for Britain.

The Melbourne railway reached Echuca on the Murray River southwest of Wagga in 1864. Bullock teams and Murrumbidgee steamers delivered Riverina produce to Echuca for swift transport to Melbourne markets and export terminals. *Victoria*, a steamboat owned by the Wagga Wagga Steam Navigation Company, regularly hauled five hundred bales of wool from Wagga to Echuca.[9] Steamers returned to Murrumbidgee River wharves with fencing wire, galvanised iron, glass windows, crockery, and other fragile and unwieldy goods.

In 1875, a Wagga newspaper published a series of letters discussing local trade and transport.[10] The male writer, an anonymous traveller, observed most business flowing south to Melbourne and back again. Storekeepers and graziers along the Murrumbidgee, he noted, received merchandise shipped from the Victorian railhead at Echuca on large steamboats and barges belonging to wealthy merchants and on smaller vessels owned by their captains. Steamers returned to Echuca towing barges laden with wool, oats, hay and other primary produce destined for Melbourne. Sydney markets prevailed north of Gundagai. Coaches and wagons climbed eastward onto the southern tablelands to meet the Sydney railway at Goulburn.

The correspondent journeyed down the Murrumbidgee from Wagga on the *Victoria*. He saw workers dragging fallen trees and branches from the river channel. Snags hindered the safe passage of steamboats and wool barges. 'It has been computed', the traveller wrote, 'that it will take ten years to remove all the snags between Hay and Narrandera.' Passing Berembed station, upstream from Narrandera, he described forests ringbarked by selectors. The *Victoria* stopped to load firewood below Murdering Island, where three decades previously squatters had trapped and shot dead up to seventy Ngarrangdhuray men, women and children.[11] On a riverbank near the island, a spacious building erected by Buckingbong squatter Frank Jenkins held bales of wool waiting for shipment.

Inland river transport had its problems. The Murrumbidgee lay particularly low during the dry year of 1872. In one month, after good rain in the river catchment, six steamboats reached Wagga. For the rest of the year Wagga residents waited in vain.

Despite the irregularity of river services, Sydney merchants and politicians worried about the loss of southern produce to Melbourne markets. To capture the southwest slopes for Sydney, workers pushed the Great Southern Railway west and south from the southern tablelands towards the Riverina.

West of Binalong, Murrumburrah residents celebrated the arrival of the railway in March 1877. They gathered at the new station, a few kilometres east of the settlement. Alongside the station, Murrumburrah's twin town, Harden, eventually grew to dominate local commerce. With the arrival of the railway, Murrumburrah district residents and produce could now travel with relative speed east to Binalong and, within hours, reach Sydney. Richard Roberts, a Murrumburrah district pastoralist, chaired a lunch at the Commercial Hotel to celebrate the opening of the route.[12] After dinner, Roberts proposed toasts to 'The Queen' and 'The Governor'. James Watson, the local member in the Legislative Assembly, toasted 'Prosperity to the district'. In the goods shed, a brass band played for hundreds of workers and their families assembled for lunch.[13]

Railway construction teams worked west to Wallendbeen, then southwest towards Cootamundra. Two steam engines hauling ten passenger carriages came to rest beside a new timber and iron railway station on the eastern side of Cootamundra one afternoon in November 1877. On rounded granite hills above the town, spearwood wattles and red stringybarks absorbed the scream of a steam train whistle for the first time. Gundagai musicians 'discoursed animating airs' as politicians and other dignitaries stepped onto the crowded platform.[14] Renowned Sydney caterers Compagnoni & Co. served a banquet to district residents and train passengers in the Albion Hotel, where a giant Union Jack decorated one wall. After rounds of toasts at the end of dinner, Minister for Public Works John Lackey rose to speak. Lackey, a farmer and publican, brimmed with enthusiasm for the Great Southern Railway, closer settlement legislation and the spread of agriculture:

> Thank you all for your enthusiastic reception today and for the cordial manner in which you drank to the health of the Ministry. I have no hesitation in saying that all other extensions to this railway sink into insignificance when compared with that which we are now meeting to celebrate (cheers). It is not necessary for me to go into particulars to show the importance of the event of today, for everyone who sees your beautiful district must be struck with its soil ('hear hear'). When one recollects that before the passing of the Land Act such places as this gave homes only to a few stragglers (applause), it is a great source of rejoicing to come now and find it sprinkled with happy homesteads and farms literally waving with wealth (cheers). I am sure that these evidences of prosperity in the future will be realised to the full. I believe this railway will be not alone a benefit to the district, but to the colony as a whole. The railway has tapped the best wheat-growing country in the colony (cheers), which will materially add to its wealth. You might have many administrations, but you will never get one with a more earnest desire to extend the railways. We are now getting into what might be called in truth the gardens of New South Wales. These agricultural districts, with their rich, fertile soils, are the mainstays of the colony's wealth. Thank you again for your warm reception. God speed the plough! (loud cheers).[15]

Eight months later, a daughter of Joseph Leary, Member for Murrumbidgee, smashed a bottle of champagne across a train carriage to open the extension of the Great Southern Railway south from Cootamundra to Junee.[16] The Wagga Wagga Steam Navigation Company foresaw an end to Murrumbidgee River transport and sold the *Victoria*.[17]

In September 1878 rail passengers disembarked for the first time beside Murrumbidgee floodplains and river red gums a few miles north of Wagga.[18] North Wagga flourished for a year as workers built a timber railway viaduct across the plain to allow steam trains passage above floodwaters. Later, trains crossed the river channel over a sturdy timber bridge.

Railway tracks linked Wagga to Albury on the Murray River in February 1881. A branch line west from Junee to Narrandera opened the same month. Sydney merchants and politicians hoped the route westward from Junee would steal western Riverina commerce from Melbourne, and along the way turn grazing paddocks into farmland.

The network of railway lines kept spreading. A decade after trains first reached Albury and Narrandera, workers laid steel tracks across hardwood sleepers from Cootamundra northwest to Temora. The *Town and Country Journal* thought the Cootamundra to Temora line 'destined to play an important part in the future development of the interior'.[19] The railway would capture

> a region remarkable for its native beauty and richness of its agricultural and mineral resources, and yet lying dormant awaiting the skill and industry of settlement.

Great ironbark forests near Temora provided timber for railway bridges and sleepers. Rich soils and good rainfall promised bountiful crops. 'It needs but little faculty of imagination', the *Journal* declared, 'to hear in the near future the hum of the stripper and thresher'. Splitting wide grazing estates into small farms would 'make the land yield as much increase as nature intended it should.'

While railways allowed efficient transport of heavy grains to urban markets, obstacles remained to agricultural development on the southwest slopes. Labour was scarce and expensive, especially at harvest time. 'Successful farming in this colony, where labour is so expensive, is beyond question without the aid of mechanical appliances', the *Cootamundra Herald* remarked in 1877.[20] Farmers turned hopefully towards British and North American inventions delivered inland by rail. At the Wagga show in 1878, a new machine able to reap cereal crops and bind sheaves with thin wire 'elicited considerable astonishment'. The device, it was claimed, did the work of forty labourers.[21] The previous year, just weeks after the opening of the Great Southern Railway extension to Cootamundra, a Canadian reaper and binder had rolled onto the station platform.[22] In a nearby paddock, locals witnessed the device fell a crop of oats and automatically gather the stalks into sheaves. Agents sold three that day, and a Cootamundra storekeeper agreed to stock the new machine. In the 1880s there appeared horse-drawn harvesters able to strip, thresh and winnow cereal crops in one operation. Farmers exchanged single-furrow ploughs for wider instruments with three or four blades. Swiftly now, cultivated spaces replaced diverse communities of grassland plants.

The railways brought profound change. In the Cootamundra region 'wheat is king', wrote a reporter in 1907.[23] In the decade following Federation, settlers yearned for evidence of national consolidation and worth. The arrival of the Great Southern Railway in the late 1870s, the same journalist observed, had transformed the district 'from a thinly-populated pastoral centre into a prosperous farming locality.' As well as growing great quantities of wheat, the region still carried as many livestock as before. For years Cootamundra had remained a 'sleeping beauty of the south'. The railway had woken her. An 'unpretentious wayside hamlet' became

> a fine town ... enjoying the advantages of an enlightened civilisation. Its streets are wide and well-kept, its buildings handsome, and its stability undoubted.

William Corby selected land north of Cootamundra early in the 1870s. A forest of white box, apple box and red gum originally cloaked the selection. 'Now there are 1500 acres on which scarcely a tree stands', a newspaper reported in 1909, 'and 1300 of them are growing promising crops of wheat and oats.'[24] Another selector north of Cootamundra near Stockinbingal farmed land once 'heavily timbered'. Now cleared, 'one can look across hundreds of acres of grand wheat land without seeing the sign of a tree or a stump upon it.'

The railway likewise enabled Temora to prosper. Ordered domestic gardens and substantial public buildings gave the town an established air. 'Five and twenty years ago the site on which it stands was covered with a dense forest', the *Town and Country Journal* explained in 1909,

> and at best the settlement was only a collection of canvas tents and bark huts. Practically the town has been erected during the last two decades, and it is now one of the most flourishing in the State.[25]

The Cootamundra to Temora railway 'pierced the isolation' of the region between the Murrumbidgee and Lachlan rivers, local historian Rob Webster wrote in 1950, 'like an artery spreading the red blood of trade and commerce'. In the service of distant markets and the imperatives of empire, railway workers and farmers dramatically altered the land:

> Gangs of sleeper-cutters scoured the ironbark ridges and swung their broadaxes over the fallen hardwood giants. The big construction camps employed hundreds of navvies and engineers, and carters and teamsters reaped a harvest keeping supplies up to them, while the farmers worked from daylight to dark on their plow-teams, then tended their clearing fires in the yankee-grubbed paddocks so that they could grow more for the bewildering but welcome market.[26]

~ ~ ~

(66) *heartland*

'FOR EXPORT', reads a red triangular stamp thumped across the image. Driving through a ripe wheat crop, a farmer in a collared shirt and a broad hat holds the steering wheel of a tractor with two hands. The tractor drags a mechanical harvester, invisible beyond the picture. An eye to the ground, the driver avoids stones and fallen branches that might damage the complex equipment. All standing grain is collected, with care and precision.

The scene is mythical, iconic of settler Australia – a productive rural vista turned golden by summer heat. The tractor itself, with a kookaburra motif fixed to its grille, is an expression of nationalism. One eucalypt stands alone, rising tall above the paddock. No young trees, wattles, hopbushes or native grasses are visible. There are no other people. Along the horizon, a sparse line of old eucalypts follows the curve of a hill. Apart from the farmer, the wheat crop and the mature trees, the land is empty.

Detail from a Sanitarium Health Food Company advertisement published in the 1964 annual edition of *Australia to-day*.

In the 1960s, this image of a wheat farmer reaping his crop was published by the Sanitarium Health Food Company to advertise a range of breakfast cereals in glossy magazines.[27] Consumers in other continents, the ad emphasised, chose foods processed and exported by Sanitarium. The advertisement located the primary meaning and significance of the paddock elsewhere. Value arose inside a relationship of local production and distant consumption, not from the paddock itself.

Dangerous dynamics operate within export-oriented systems of agricultural production, marketing and consumption. Globalised market forces block the attentiveness of consumers and producers to the shifting and particular ecological needs of rural places. People are pressured to respond to strident market signals before the subtle and complex messages of rural places and of other species. 'Modern producers and consumers', philosopher Freya Mathews explains,

> decide what will be produced and what will be consumed well in advance of discovering what is actually on offer, what is already available in their local environment. They are *makers* rather than *finders*, preferring their own version of reality to reality as it is given.[28]

Geographic and cultural divides between city and country maintain demand for uniform food products, not diverse, seasonal produce. Urban dwellers cannot easily come to know and care for particular farmers or the natural patterns of rural places, nor begin to see how production and consumption may be folded into local ecologies. Denial of connectivity and the impeding of responsive relationships promotes the ecological disorders present today across Australian farming regions.

Since the onset of colonisation, the operation of global trading networks has shaped perceptions of Australian rural places. In 1848, English philosopher and economist John Stuart Mill described the British colonies as

> hardly to be looked upon as countries, carrying on an exchange of commodities with other countries, but more properly as outlying agricultural or manufacturing estates belonging to a larger community.[29]

Production and export of primary produce gave Australian settlers purpose and identity within the British Empire. According to popular agrarian mythology, rural land should be populated by many farming families, honest and stable, not a handful of wealthy pastoralists. In exchange for exported manufactured goods, Britain absorbed surplus Australian primary production.

Like any story, the Australian agrarian myth served certain interests. The population of England, Scotland and Wales exploded over the nineteenth century, tripling to forty million.[30] Centuries of industrialisation and urban expansion made Britain largely dependent on food produced elsewhere. At the start of the twentieth century, the new Commonwealth of Australia sought firm national foundations in farming communities, conservative and prosperous, generating food and fibre for export.

In 1904 Massey-Harris, a British and Canadian farm machinery manufacturing firm, published 'The Breaking of the Drought'.[31] The cheerful song-advertisement reflected and reinforced Australian agrarian mythology, and the dynamics of nation and empire the mythology served. Good rain has ended a miserable drought, and a bumper wheat harvest is on the way:

> Things are merry in the country, where a year or two ago
> You never met a farmer but you heard a tale of woe
> For the days of drought are over, days of dust and sand and heat,
> And the summer once again will see the harvest of the wheat.
> Wheat! Wheat! That picks a broken country up and sets it on its feet.

The value of wheat ripening in Australian paddocks lies elsewhere, in faraway London:

> The man who has a bit of land forgets his troubles old,
> When day by day he sees the green grow ripening into gold.
> And round his happy homestead there was never sound more sweet,

> Than the whisper of the breezes as they rustle through the
> wheat.
> Wheat! Wheat! That will turn to golden guineas when the
> London bankers meet.

Money waiting in London promises to erase hardship from the wheat farms of Australia:

> There is schooling for the kiddies, there's a present for the wife,
> And a cheque to pay the doctor for the time he saved her life;
> There is cash to meet the past due bill that almost had him
> beat,
> And a little fixed deposit in those waving fields of wheat.
> Wheat! Wheat! He will face his banker boldly when the
> harvest is complete.

A global trading network delivers this happy turnaround in local fortunes. Exchanged for Australian wheat, British and Canadian machines are sent inland by rail:

> The Massey-Harris binders go on ev'ry railway train,
> And their Harvesters are rolling out to reap the golden
> grain.
> All the ports are getting ready to accommodate the fleet
> That is heading for Australia just to take away her wheat.
> Wheat! Wheat! Five million golden acres for the world to
> buy and eat!

~ ~ ~

In spring I take a rough road past old apple box trees. They stand alone in paddocks glowing yellow with flowering canola. My car heads down the pathway to a creek, then rises again. Connaughtmans Creek curves around a hillside where the road meets double gates of weathered iron and rabbit netting. Parking the car, I notice an ancient poplar tree nearby, stocky and venerable. A brown bottle discarded decades ago lies beside native geranium on a granite outcrop. I unhook the gate chain and walk into Hardies Reserve.

The track into the travelling stock reserve levels out beside Connaughtmans Creek. Walking, I find a yellow and brown fragment of snake-necked turtle shell on the sandy path. Further along, a stone flake with a sharp edge is pressed into the road. My finger traces the bulb rising from the grey, fine-grained material, evidence of a hammer-stone striking the flake from a larger core. I leave the track and climb a gentle slope past tussocks of spear grass towards yellow box saplings. A chocolate lily flower stem sways near fallen branches. Dark and feathery clumps of *Acacia deanei* rise from boulders. I walk down to the waterway, past sheep bones and flecks of wool, into the shade of an old weeping willow. Fresh leaves emerge on pendulous branches. Frogs call and small birds dart in a dry stand of cumbungi.

The primary local pathway of settler trade once crossed Connaughtmans Creek here, inside Hardies Reserve. Dark pines in a paddock to the west mark the site of Wallendbeen station's first homestead, a simple building of eucalypt slab walls. Traffic from the south and west converged at the headquarters of the pastoral station and continued eastward through Murrumburrah, onto the southern tablelands, then to Sydney. Teamsters drove bullock wagons along the road, delivering wool to Sydney and collecting supplies.

When the Great Southern Railway arrived in the late 1870s, Wallendbeen village grew northwest of the station homestead. Surveyors marked out a new road along a ridgeline north of Connaughtmans Creek.

The New South Wales government set aside Hardies Reserve for travelling stock in 1883. Drovers with sheep, cattle and horses periodically camped in the reserve beside the creek. Unlike most inland slopes, travelling stock reserves never experienced repeated cultivation, continuous grazing and artificial fertilisation. Long rests often followed days of intense grazing pressure, perhaps with similar effect to Wiradjuri burning practices. Inside travelling stock reserves, a diversity of plants continued to flower, set seed, germinate and grow.

I leave the cool shade of the willow tree and head upstream to where the track crosses the narrow waterway. On the western bank, weathered timber beams protrude from dark earth. The rotten pieces of wood look like the remains of an early road bridge. On the other side of the creek, I notice many stones carted and dumped, a solid path for heavy wagons and mobs of sheep. A wide network of shallow gullies lies across the hillside to the east. Some are active, others grassed over. The roadway has shifted many times, it appears, as rainfall channelled and swept earth away.

I walk uphill across the eroded surface towards a dip in the skyline where the abandoned road descends on the other side. Dark, glossy leaves of apple box trees shift and glisten in the afternoon breeze. The drawn whistle of a diesel train entering Wallendbeen comes from the west, then a rumble of carriages heavy with grain or shipping containers. A small brown hawk appears above me, circling and chattering, narrow wings beating swiftly. I have disturbed it, and my presence is resented. For a moment the predatory bird hovers above a tall jumble of split boulders, encrusted with lichen. I gaze into the shadowy spaces between the stones, and wonder what early settlers imagined as they passed here, encountering these slopes for the first time. From the east, along the pathway through Hardies Reserve, came hopeful, restless colonists. In the other direction, to the port city of Sydney, went gold nuggets, bales of wool, herds of cattle and bags of wheat.

Francis Sweeny journeyed westward along the road towards Wallendbeen in 1870 with Irwin Smith and James Morrow. The men grew up together in rural Northern Ireland. In Australia, they carved dairy farms from coastal forest south of Sydney. Years later, when policymakers encouraged closer settlement inland, the three young men headed west across the Great Dividing Range. Near Goulburn, a bullock wagon driver advised the Irishmen to keep going until they reached Connaughtmans Creek, beyond the village of Murrumburrah. They selected land on Wallendbeen station, then returned east to collect their families.[32]

The stony slopes of Hardies Reserve were once part of Allowrie, a farm established by Irwin Smith and his family. Church of England ministers delivered sermons inside the Smith family homestead. In 1876, parishioners built a church from eucalypt slabs on the property, beside the main road on land now inside the travelling stock reserve. When the railway came, they moved the church to a new site in Wallendbeen village.

James Morrow and Francis Sweeny selected land at Cullinga, a few kilometres south of Allowrie. As an old man, Francis Sweeny's son Frank told his niece about the early days on Brae Farm, the Sweeny family selection. Before his father built fences, Frank Sweeny said, bullocks would disappear into expanses of tall kangaroo grass. Francis strapped bells to their necks to save trouble searching for them. On a low stretch of land, the selector sank a well and found fresh water six feet down.

Francis Sweeny ringbarked box trees and red gums with a sharp axe. He lit fires in pits dug below each trunk. Falling branches shattered as the dead trees burned. He gathered fallen wood and set about burning down other ringbarked trees. Having cleared the eucalypts, he enclosed the new cropping paddock with a brush fence made from smaller trees. Cullinga district farmers first planted maize. Between the rows they grew pumpkins, watermelons and pie melons. The following season the new selectors sowed paddocks to wheat.

When Francis first cultivated a paddock on Brae Farm, taproots and tubers – his son Frank called them 'wild carrot roots' – accumulated on his single-furrow plough, temporarily stopping its progress. Many grassy woodland forbs grow fleshy roots. Taproots and tubers enable the plants to survive fires and droughts.

Before colonisation, Wiradjuri dug into the red basalt soil of Wallendbeen and Cullinga to harvest roots and tubers. Yam daisies, milkmaids, chocolate lilies, early nancy, native geranium and other species offered a diverse selection. Francis Sweeny had trouble removing the many roots and tubers gathered on his plough. The problem did not recur the following season. Tuberous grassy woodland plants no longer grew inside the paddock on Brae Farm.

Demanding production

One spring night in 1924, Sir Dudley and Lady de Chair journeyed southwest from Sydney in a vice-regal carriage hitched to the Albury mail train.[33] Morning sunlight caught the train passing green crops of wheat and oats. Sir Dudley, governor of New South Wales, arrived in Cootamundra to open the annual show. Inside showground pavilions, judges inspected merino wool, butter, lucerne hay, wheat and barley.[34]

Sir Dudley and Lady de Chair toured the district. They watched sheep shearing at Glen Iris station, near Bethungra. At the Cootamundra Town Hall, an ornate Victorian building, civic leaders gave the governor an illuminated address, a rectangle of cardboard adorned with watercolour roses, honeysuckle, pansies and gilded scrolls.[35] Fancy inscriptions declared loyalty to 'the Throne and Person' of King George. In the recent European war, the presentation reads, 'exceedingly large numbers' of Cootamundra district men fought for the British Empire. The address offered Sir Dudley 'a sincere and hearty welcome to this one of the leading producing centres of the State'.

When Sir Dudley and Lady de Chair visited Cootamundra the Catholic Church on Morris Street was almost new. Looming above the town, the voluminous red brick building remained for seventy years the most prominent structure in Cootamundra. Today, when people climb one of the stony hills beside Cootamundra, they see three giant white silos dwarfing streets, houses and the church. At the silo facility beside the railway line, machinery lifts grain skyward before sending it down chutes to tumble through dark storage space. Between the steel monoliths, pigeons flap and disappear into shadows. Wheat trucks roll across quarried stone to deliver produce from local paddocks. Heavy trains rumble away, hauling the harvest to urban granaries and coastal export terminals.

Workers began constructing the new silo complex beside the railway line in the winter of 1981. The central elevating tower would rise more than fifty metres, the *Cootamundra Herald* marvelled, twice as high as the post office clock tower. According to the newspaper, the new complex promised 'to be more than just a tourist attraction and a talking point – it is to serve primary production.'[36]

GrainCorp issues stylised maps of the grain storage and delivery infrastructure it controls. The maps show names and locations of silos on a network of railway lines linking the western slopes and plains of New South Wales to coastal centres. Blue lines traverse the southwest slopes. Silo complexes beside towns and paddocks are marked and named in succession. One rail and silo pathway runs east from the western margins towards Junee: Narrandera–Grong Grong–Matong–Ganmain–Brushwood–Coolamon–Marrar–Old Junee. Lines converge at a black circle signifying the giant silos beside the railway line at Cootamundra. A single blue path runs east to the coast – fast and straight as a farm auger – to the export terminal south of Sydney at Port Kembla, a larger black circle on the map. Green lines gather on the central slopes and head east to Sydney. From the north, red and orange pathways feed south and east towards a black point denoting Newcastle, a port city north of Sydney.

The coloured pathways look like great river deltas in reverse – channels across a wide region drawing moisture and nutrients together and away. Trains deliver vast quantities of grain to central points of consumption and export. In 2004 almost eighteen million tons of wheat – most of the national annual harvest of about twenty million tons – left Australian shores on cargo ships.[37]

Jack Hallam, New South Wales Minister for Agriculture, arrived in Cootamundra for the official opening of the new silo complex in September 1982. A photo in the *Cootamundra Herald* shows men posing as Hallam uncovers a brass plaque. Farmers drove away from bare, drought-stricken paddocks to join the ceremony and look over the silo complex.

In the early 1960s, local farmer and Cootamundra Silo Committee chairman Ross Faulks told the crowd, less than four thousand tons of wheat entered the Cootamundra railway silos at harvest time. Trucks now delivered about forty thousand tons to the railway storages each summer. District farmers had badly needed a much larger silo facility. The new complex took grain twice as fast – up to four hundred tons per hour. Each of the three welded-steel containers stored ten thousand tons of grain. Operators used remote control technology. Transferring grain between trucks, silos and railway carriages required only two workers.

Minister Jack Hallam said his government was determined to minimise costs and foster economic development. The particularly dry and dusty hillsides beyond the town did not suppress his optimism. He told the crowd:

> I am sure that continuing developments in wheat production will result in district producers utilising this valuable facility to the fullest extent in the future.[38]

At the end of World War II, Australian wheat farmers were harvesting about half a ton of wheat from each hectare of crop. By 1990, average yields had tripled.[39] In the face of declining terms of trade and strengthening global competition, farmers adopted new plant varieties and agricultural technologies to push more produce from paddocks.

Pressure from global trading markets to maximise production corresponded with official economic agendas. Policymakers abandoned old notions implicit in land legislation since the introduction of the Robertson Land Acts in 1861. Family farming was no longer honoured and protected as a stable foundation for society and nation. As words like 'deregulation', 'privatisation', 'Reaganomics', 'Thatcherism' and 'transnational' entered the Australian lexicon in the 1980s, people began to see rural land primarily as the material basis of a competitive global system of industrial production and trade. Australian bureaucrats worked within international teams to free agricultural commodity trade from economic barriers thought to promote inefficiency.

According to dominant opinion, Australian economic competitiveness depended on the aggressive promotion of free trade. In 1991 the federal government announced a timetable for the dismantling of all protective tariffs and quotas.[40] For generations, producer cooperatives and statutory marketing authorities had given farmers and farmland some protection against the volatility of global markets. Powerful advocates of strident, neo-liberal economic theories dismantled or corporatised these protective organisations. Dissenters were few. Even the National Farmers' Federation supported the free trade agenda. In Australia and across the globe, as Indian activist Vandana Shiva observes, economic fundamentalism turned 'protection', a term signifying care and respect, into 'a dirty word, the worst crime of the global market place.'[41]

Owen Whitaker's ancestors began farming in the Mitta Mitta district, east of Junee, late in the nineteenth century. He described the immense risks farmers today must take as they strive to generate vast quantities of primary produce. Generations ago, Owen explained, farm business risks were generally smaller, and were shared among the local community. When things were tough on the farm, shop owners rarely demanded immediate payment. Farming families grew most of their own food and were less dependent on external networks. When drought struck or prices collapsed, his grandparents may have lost a few hundred pounds, but were able to carry on. 'Nowadays it's a huge blow', Owen said, 'and you've got the bank chasing you, and you've got compounding interest rates, and refinancing and all the rest of it.' The high-input, high-output farming model is characterised by great risks and costs:

> There's bigger stakes, much bigger stakes involved in agriculture today. This situation has to affect the people and the landscape much more. Something's got to pay for that. As you've got bigger inputs, you've got more and more people disconnected with the farm that you've got to pay every year. And there's only two ways you're going to pay them. One is through increased net income from production, and the other way is with finance. Sooner or later the land has got to pay, it's got to get that extra return to allow for the ever-increasing inputs. And that comes as a very negative force with strong imperatives driving higher production against the future best interests of the land or its people.

Forces propelling and shaping production systems emanate from the global marketplace, a domain in which individual farmers are powerless. While farmers must respond to economic pressures, globalised market structures are unresponsive to the social and ecological needs of particular agricultural regions. High debt levels and production imperatives constrain the ability of many farm managers to care for land: 'It's hard to be green when you're in the red', farmers say.

Owen took over the family farm at Mitta Mitta from his parents in the late 1970s. 'Don't overcapitalise, don't push your land too hard, rely more on its natural production attributes', his father told him. Looking back over his decades as a farmer, he saw the wisdom of his father's advice:

> Don't kill the greatest wealth of the land – its ability to produce naturally. Because, that's the only thing that can keep you independent. And that's a great thing about being a farmer – to retain your independence. And I believe that we're losing that. We're losing it so

severely. And what have you got then? You're living in a community with a degraded landscape, with a degraded infrastructure. That means that your kids don't want to be around or they can't afford to be around. And if you lose that, and if you lose your independence to be able to produce in your own right, using a natural nutrient-cycling system, really what have you got? You've got a business that produces a commodity in a climate that is very erratic, producing into a market that is dictated by overseas interests that you have no control of, in a production system that you've lost control of because you have to buy an ever-increasing amount of inputs.

In the 1980s and 1990s, as Australian agriculture became increasingly exposed to the insistent and inflexible demands of the global marketplace, an ideology of 'productionism' began to dominate farming life. Ethicist Paul Thompson defines productionism as 'the philosophy that emerges when production is taken to be the sole norm for ethically evaluating agriculture'.[42] Australian sociologist Geoffrey Lawrence considers the term to mean

> the system of agriculture in which efficiency and productivity goals are privileged over environmental and community-based desires and concerns.[43]

According to Lawrence, productionism 'justifies the use of "high tech" solutions to problems, and tends to place farmers on a technological treadmill.'

Representations of Australian rural places almost exclusively in terms of production reflect and maintain a dominant culture of productionism. On its website, Harden Shire Council defines the local district as a centre of primary production:

> Wheat, canola, stone fruit, wool, wine, mustard oil, lamb, beef and livestock are but a few of the products from the area affectionately known as Australia's Produce Supermarket.

At Wallendbeen, twenty kilometres west of Harden, sculptures stand beside a busy intersection. 'CELEBRATING THE AUSTRALIAN WHEAT INDUSTRY' declares a timber sign. Concrete pillars resembling wheat stalks support acrylic panels, each a stylised head of wheat. Fibre cables cast shifting light through the panels. Green gives way to gold, suggesting the movement and ripening, a brass plaque located in Cootamundra explains, 'of a dense crop of wheat.' Further south, drivers entering Junee Shire pass a sign decorated with official district motifs: one fat lamb, three sheaves of wheat, two bunches of grapes and a train engine.

Productionist representations of rural places reaffirm a destructive dynamic. As forester and biologist Aldo Leopold argued, seeing land predominantly in commercial terms leads to the neglect and elimination of elements lacking immediate economic value.[44]

Assumptions that the economic components of farmland would continue functioning without the presence of the uneconomic parts has underlain the development of Australian agriculture. Until recently, the ecological blindness of industrial farming pervaded Australian tax law. For decades before legislative changes in 1983, primary producers were not required to pay tax on income used in

> the draining of swamp or low-lying lands where that operation improves the agricultural or grazing value of the land [nor in] the destruction and removal of timber, scrub or undergrowth indigenous to the land.[45]

Drivers nearing Harden from the east pass grain storages beside the railway line. Entering the town they read: 'HARDEN, CENTRE OF THE BEST WHEAT GROWING LAND IN AUSTRALIA'. Four ripe, grain-filled heads of wheat decorate the celebratory metal sign.

The abundant yields often delivered by industrial farming suggest that all is right and well in modern agriculture. Yet an emphasis on bounteous production obscures ecological costs. Ecological scientists recently rated Jugiong Creek, a major waterway draining the Harden district, the most degraded stream in the entire Murrumbidgee River catchment.[46] Locally, soil acidification and dryland salinisation are mounting problems. If we look beyond production and consider ecological stability and wellbeing, the Harden district appears particularly unsuited to the broadscale, industrial style of wheat production practised there today.

Cunningar railway silos, Harden district.

In the last decades of the twentieth century, productionist agriculture generated an array of ecological problems on the southwest slopes. Responding to patterns of disorder, government agencies and agribusiness groups unfortunately reaffirmed productionist imperatives.

Few people questioned the appropriateness of high-input, high-output, export-oriented farming, a production model diverting attention away from the ecological needs of rural regions. Instead, responses tended to reinforce established relations of human mastery over rural places. Farmland remained commonly perceived as a 'natural resource' available for industrial use within global systems of production and trade.

The dominant framework of modern farming abstracts people from natural systems, positioning humanity above and in control over rural terrain and other species. In 1999, the federal government published a discussion paper called *Managing natural resources in rural Australia for a sustainable future*. Government agencies prepared the paper in collaboration with landholders, scientists, agribusiness representatives and conservationists. It lists a set of principles to guide the formation of national policy. 'Ecologically sustainable development', the first principle, reflects a widespread belief that the function of natural systems is to serve people and their economies:

> Ecologically sustainable development – which involves maintaining and enhancing healthy ecosystems and biodiversity and using resources soundly for continuing wealth creation to meet social aspirations – is the framework for managing our natural resources, now and in the future.[47]

Declining terms of trade and intense global competition set in motion destructive dynamics. Farm managers applying mainstream, industrial production methods use great amounts of artificial fertilisers, herbicides and insecticides to make rural places produce vast quantities of output. According to Mark Harris, a Wagga farm management consultant, broadacre farmers must follow three simple rules if they are to remain profitable: 'know what you do well, produce as much of it as possible, and produce it as cost effectively as possible.'[48] Attention to components of the land seen as not directly contributing to productivity and profit does not, in the short term, make business sense. As governments dismantled social security systems late in the twentieth century, people failing to compete economically faced painful consequences.

Desperation to maintain high and rising levels of production drove the emergence of the Landcare movement in the 1980s and 1990s. For decades, narrow attention towards production had obscured the ecological consequences of industrial farming practices. Government agencies and land managers took a wider perspective as they realised that mounting ecological disorders were undermining agricultural productivity. Rather than rejecting productionist agendas, however, the Landcare movement tended to reinforce dominant and ecologically destructive cultural frameworks. As a group of Australian Bureau of Agricultural and Resource Economics economists explained, the primary role of Landcare was to erase another barrier to the ever-rising productivity of industrial agriculture:

> Efforts to achieve gains in productivity are increasingly constrained by the inherent productivity of the natural resources available. Over the past decade, the Landcare movement has been the focus of a nation-wide attack on land resource problems.[49]

Government agencies, farmers and Landcare groups hoped that minor interventions would allow broadacre farming practices to continue relatively unchanged. They planted native trees and shrubs across salty and eroded areas fenced from stock and in bands along paddock edges. The primary aim was not to relieve rural places of pressure. Rather, scientists and Landcare groups hoped to help farmland bear the ongoing demands of high-input, high-output, export-oriented agriculture. People seeking to address the ecological disordering of rural places rarely acknowledged and challenged the economic and cultural dynamics responsible.

Ignoring the underlying origins of ecological problems gave minimal safeguards against future damage. In *Productive use of salt affected land*, a pamphlet published in 1993, NSW Agriculture suggested no changes to farming methods causing dryland salinisation. Only symptoms of the disorder needed addressing, the publication made clear:

> Reversing the *causes* of dryland salinity by treating the catchment is generally a difficult and slow process. Treating the *symptoms* of dryland salinity is an important part of an overall catchment or farm plan. This involves revegetating saline discharge areas with more productive species.

Efforts to reap profit from situations of disorder reinforce the narrow commercial imperatives underlying problems like dryland salinisation. The Productive Use and Rehabilitation of Saline Lands (PUR$L) group of Australian natural resource managers and scientists meet annually to discuss 'a different vision' of salinisation: 'saltland is also potentially useful for profitable industries in agriculture, forestry, horticulture, aquaculture, minerals and energy.' In 2001 the New South Wales Department of State and Regional Development launched 'SalinityBiz', an enterprise designed to bolster business activity and primary production in regions where farming practices caused salinisation. Program architects hoped SalinityBiz would help people overcome limits to production and profit imposed by dryland salinity. The 'major challenge' and 'growing burden' of salinity, they declared, also represented 'a tremendous opportunity for people and organisations with the skills and resources to develop and implement solutions.'

According to the SalinityBiz website, salinity could be viewed 'as a business opportunity as well as an environmental scourge'. Efforts to find 'solutions' to the problem of salinity would 'generate jobs, investment, sales and exports.' Within the SalinityBiz program and like-minded responses, the cultural and economic drives responsible for salinisation are invoked to counter the problem.

Before the intensification of colonisation on the southwest slopes in the second half of the nineteenth century, Wiradjuri people continued tending land to bolster ecological connectivity and heighten natural productivity. Ecological dynamism is a feature of healthy and diverse natural communities. Interactive natural relationships offer land resilience against drought, storms, frost, wind, bushfire and other forceful events. Natural stability and productivity is retained.

The extreme demands of global markets ensure that farmers maintain industrial production systems rather than integrating agriculture into local ecologies. Ecological integrity and wellbeing, explained Owen Whitaker, have become irrelevant in the scale of things that you have to achieve, to be able to produce economically, to return the profits. And they've become irrelevant in the world scene of marketing. Because they say: 'Look, if you need to look after your land to that extent, that's fine, if you want to do that. But, the world market's here to say you're not relevant in the scheme of things, in the world market. Because there's producers in other countries that don't really care about that and they'll produce and compete with you on an equal footing on a world scale. Because we've got cheap fossil fuels that can move all this sort of stuff around, and we don't have industrial relations laws in those other countries, so we can exploit our workforce. Or we don't have environmental laws or regulations so we can exploit our environment.' And if you want to be a producer on a world scale you're got to swallow that, and compete.

Today, efforts to make industrial, high-output agriculture 'sustainable' rarely reflect any fundamental shift in values and beliefs. Unrelenting global economic pressures and entrenched cultural tendencies block necessary change.

In modern farming, the rhetoric of 'sustainability' reaffirms rather than challenges the culture of industrial productionism. According to the *Macquarie dictionary of new words*, the term 'sustainable farming' entered Australian English in the 1980s.[50] Definitions of 'sustain' in *The Oxford English dictionary* include to 'keep going, keep up, ... to carry on (a conflict, contest)', to 'maintain the use, exercise, or occupation of', to 'endure without failing or giving way; to bear up against, withstand', and to 'bear, support, withstand (a weight or pressure).' The dictionary defines 'sustainable' as 'capable of being borne or endured', and 'capable of being maintained at a certain rate or level.' And 'sustained' is defined as:

> kept up without intermission or flagging; maintained through successive stages or over a long period; kept up or maintained at a uniform (esp. a high) pitch or level.[51]

Comments made in 2003 by Ian Macdonald, New South Wales Minister for Agriculture and Fisheries, show the relationship between notions of 'sustainability' and the dominant culture of agricultural productionism.[52] 'Agriculture has never been a more critical sector of this state's and this country's economy', Macdonald declared. The annual value of agricultural production in New South Wales, the politician explained, was almost nine billion dollars, nearly one third of the national agricultural output. Such a figure was achieved through steady productivity growth since the 1970s, a result of better technologies and land use practices:

> Continual improvements in production technology and techniques have allowed more effective use of labour, land, water and nutrient resources.

Macdonald ignored the array of ecological disorders imposed in recent decades by the postwar intensification of farming methods. Instead, he praised the 'more effective use' of farmland. In the future, according to Macdonald, farmers would have to boost production levels even further to meet the demands of a growing population. 'Sustainability' of rural land and people depends on productivity growth, on more of the same:

> Only through the continual refinement of existing technologies – and the development of new ones – will NSW agriculture be able to deliver on these needs and, at the same time, create long-term economic and environmental sustainability.

The logo of the Harden–Murrumburrah Landcare Group is a grain-filled head of wheat. In modern 'sustainable farming', profitable and increasingly bounteous production remains the primary goal, as NSW Farmers makes clear:

> Natural resource management is an important part of everyday life on most farms, as landholders work to make their properties more productive, profitable and environmentally sustainable.[53]

Productionism infuses the Landcare movement. 'Financial returns from the land', explained agricultural scientist David Smith at a recent Landcare conference in Cootamundra, 'are very important to individuals, and to the nation. Landcare', he continued, 'must be an integrator/mediator of caring and cashing.'[54]

In 2000 Warren Truss, federal Minister for Agriculture, Fisheries and Forestry, addressed the International Landcare Conference in Melbourne. Despite the ecological disorders delivered by export-oriented, productionist agriculture, the primary considerations of modern farming remained production, profit and the national economy:

> Landcare has emerged in Australia over the past 20 years to become part of the mainstream of Australian agriculture. The direct link between long-term sustainability of the resource base and the productivity and profitability of industry is now well established and accepted. In 1998–99, Australian agricultural production is valued at around $28.6 billion and contributed $22 billion in export earnings to the Australian economy.[55]

Early in the 1980s the Department of Education exchanged our old and capacious school bus for a much smaller vehicle. The farmland west of Cootamundra held fewer children. Between 1961 and 2001, the number of Australian farms almost halved. Recently, the Australian Bureau of Agricultural and Resource Economics told farm managers they could boost profitability by purchasing or leasing even more land.[56]

With official encouragement, rural depopulation continues. As farmers depart and farm acreages expand, notes philosopher Wes Jackson, the 'eyes-to-acres' ratio falls.[57] Global dynamics of trade and economic power drive families and individuals from rural places of emotional attachment. Human memory and knowledge of local ecologies, in some cases gathered over generations, vanishes from the land. Depopulation and rising farm sizes mean that remaining farm managers are less likely to inherit and develop intimate understandings of rural places. Potential for responsive relations between people and farmland is undermined. In pursuit of profits and production, on expanding acreages, farmers cannot easily sense and meet the varied, shifting needs of local ecologies.

An overemphasis on production, observes farmer Wendell Berry, obscures the natural and cultural foundations of long-term agricultural productivity.[58] Obviously, farming must be productive. But production is not, Berry argues, the primary standard by which we should measure and shape agriculture. Enduring productivity depends on ecological and social wellbeing. Land and people must come before economics and production. For land to be strong and naturally productive, writes Berry,

> the people who use it must know it well, must be highly motivated to use it well, must know how to use it well, must have time to use it well, and must be able to afford to use it well.

Standards determined not by global markets may be defined for Australian agriculture. Social and ecological needs can be met, responsive ties between people and land enabled. Suppose the central standards of modern farming became not production and profitability, Berry proposes, 'but the health and durability of human and natural communities'.[59] What outcomes might then unfold for the farmlands and people of Australia?

progress

Stories of inevitability

When Evelyn Sturt became the first Commissioner of Crown Lands for the Murrumbidgee squatting district in 1837, he encountered vibrant, enchanting terrain:

> The country was at this time most beautiful – miles of it untrodden by stock, and, indeed, unseen by Europeans. Every creek abounded with wild fowl, and the quail sprung from the long kangaroo grass which waved to the very flaps of the saddle.[1]

Ecological decline was 'a necessary change', the commissioner later wrote, an unfortunate by-product of inevitable progress:

> It has often been a source of regret to me that all the charms attending the traversing of a new country must give way to the march of civilisation.

The 'demands of population', he observed, destroyed 'the quiet serenity of the Australian bush'.

Does the meeting of human need require extinction and disorder? Are people inescapably destructive? Like Evelyn Sturt, Mary Gilmore remembered lively swamps and grassy woodlands across the western slopes. Wiradjuri sanctuary law, burning methods and other practices ensured diversity and abundance:

> The richness of unbroken centuries of an untilled land sapped through everything. Life teemed in the water, life teemed on the

earth, life teemed in the air. Life fed upon life in balance; bred, multiplied, and knew no famine.[2]

In the 1860s and 1870s, as new closer settlement legislation and the Great Southern Railway brought the displacement of Wiradjuri people and the break up of vast squatting runs, Gilmore saw land lose vitality and productivity:

> A few years later when I asked my father why we could not get fish as formerly he said, 'When the blacks went the fish went': meaning that the habit of preserving the wild was destitute in the ordinary white settler. Yet at that time the white population on the rivers was only a fraction of what the black had been.[3]

Humans are not innately and unavoidably destructive, Gilmore and her father knew. Culture shapes action. Different perceptions and beliefs foster different behaviours and different ecological realities. Traditional understandings imported from western Europe and reinforced by powerful colonial dynamics generated the narrative of progress and inevitable ecological destruction told by Sturt. Progressive stories justified Aboriginal dispossession and the fragmentation of living systems. On the southwest slopes, the honouring and naturalisation of relentless change enabled the denial of the needs and rights of Wiradjuri people, other species and local ecologies. Colonial power and distant markets demanded faith in stories of progress.

Powerful narratives of inevitable change helped ease colonial anxieties about the natural potential of inland Australia. In 1909, a journalist visiting Temora described 'the fears and doubts of early settlers that wheat-growing would prove unstable'.[4] The reporter assured readers that such uncertainty was groundless. Farmers had survived the great drought at the end of the nineteenth century, and wheat acreages continued to grow each year.

Might a longer and deeper drought descend? After only a few decades of settlement and weather observation, colonists couldn't be sure. Might bushfires or erosion undermine the colonial project of agricultural settlement? Could the western slopes of New South Wales really help build secure, productive foundations for the emerging nation of Australia? Or did the land itself call for other imperatives, other stories, other styles of engagement? Grey stands of ringbarked eucalypts reminded British colonists of the region's distinct nature. Restless and active, settlers evaded persistent questions and shadowy doubts as they drained swampland, buried grasslands with ploughs and established permanent settlements.

Closer settlement laws and the new railway fuelled agricultural development across the Wallendbeen district in the final decades of the nineteenth century. Selectors carved small farms from the grazing paddocks of Wallendbeen station, a property managed by the Mackay family since the early 1840s.

When Alexander Mackay died in 1890, the *Cootamundra Herald* imagined the hurt felt by the pastoralist over the fragmentation of 'his magnificent estate'. More than sixty thousand acres in 1848, the squatting run had dwindled sixfold. Despite such loss of grazing land to farming families, the newspaper obituarist 'never heard of any bitterness or bickering between the selectors and the pioneer of Wallendbeen.'[5]

Alexander Mackay's son Kenneth accepted the decline of pastoralism and the rise of agriculture in his poem 'The Passing of the Shepherd Kings', published in 1908.[6] For the progressive story of nationhood to unfold, the heroic first chapter of pioneering had to close:

> Vanguard forever doomed to die!
> The hour draws near,
> When rope and shear
> Will, frayed and rotting, lie;
> When camp and yard will pass away,
> And bit and steel will useless rust
> In empty stalls,
> Where silence calls
> To silence, 'mid dishonoured dust.

Brave squatters like the poet's father had laid a firm foundation for the Australian nation and its people:

> With iron will and steadfast face
> You led the way,
> In that dim day
> Which saw the dawning of our race.

Like other legends of Australian pioneering, 'The Passing of the Shepherd Kings' honoured men and women who endured hardship as they strove to secure land for settlement and production. Squatters and other pioneers battled hostile natural forces, the legends hold, and bequeathed the spoils to many:

> Empires have cradled in thy tents;
> And millions hold, because of you,
> The lands you won,
> From snow and sun,
> When sea and shore alike were new.

Did Kenneth Mackay have in mind stories told by elderly men of brutal encounters with Wiradjuri when he wrote of squatters wetting soil with blood?

> No foot of our Australian soil,
> But you have wet
> With blood or sweat,
> And sanctified with manly toil.

Squatters lost their extensive pastoral runs for a noble cause, Mackay felt. Today, government policy and economic structures encourage farm expansion and rural depopulation. A century ago, Mackay and many other Australians valued the presence of humans in agricultural regions. Crowded city streets, they believed, undermined the moral fibre of the emerging nation. New farming communities promised social vigour and stability:

> To-day your hoof-trod lands we need,
> So all the past
> Must be recast,
> That men may garner strength to breed
> A sturdier race than fetid spawn
> In narrow streets and filthy hives,
> Where crime takes shape,
> And passions rape
> The Godhood out of human lives.

As the inevitable came to pass and closer settlement split apart wide grazing estates, a wistful Mackay declared he would never forget the heroic pioneers who first secured control over the land:

> Full soon this fair Arcadian dream
> Of primal peace
> Alas, must cease!
> For, close at hand, strange watch fires gleam,
> And keen eyes mark our empty plains.
> So men must come, and sheep must go,
> If we would hold
> This land of gold
> Our fathers won us long ago.
> But when one fat with wine and corn,
> Who has forgot,
> Or knoweth not
> The tale of how his race was born,
> (In love with his own pampered self),
> The song of farm and orchard sings —
> Whate'er his boast,
> Be mine to toast
> The memory of the Shepherd Kings.

An epic narrative of nation building and the inevitable spread of agriculture drove the fragmentation of Wallendbeen and other pastoral stations. The editor of *The Pastoral Homes of Australia* wrote in 1929:

> Where today is a vista of ploughed earth or flourishing crops and all the other visible signs of a farming community, not so very long ago stretched broad sheep pastures cleared, fenced, watered and hewn from the rough.

The relentless march of time would itself destroy the remaining grand estates of pastoral pioneers:

> Inevitably, in the natural order of things, will their work of to-day become part and parcel of the history of Australian development, for these homes, or at least the majority of them, will sooner or later no longer exist as we know them. The tide of closer settlement, a tide that will not recede, is advancing and submerging them utterly, or leaving here and there surrounded by its flood an isolated island, the homestead block of some famous old station.[7]

In his book *The New South Wales wheat frontier 1851 to 1911*, geographer Michael Robinson presents a series of maps charting the spread of farmland across the western slopes and plains.[8] Black dots, each representing one hundred acres of wheat, coalesce near railway lines. Three maps – 1892, 1902 and 1910 – show dark impressions gathering into a dense expanse across the southwest slopes.

In the late nineteenth and early twentieth centuries, agricultural development and the breaking up of pastoral leaseholds drove remnant Wiradjuri groups from station campsites. Native foods vanished as farmland spread. Managers of dwindling pastoral stations no longer needed large workforces. One account records the last resident Aborigines of Ironbong station, near Bethungra, leaving around 1910.[9] Displaced families and individuals gathered at town fringes.

The close proximity of hungry, dishevelled Wiradjuri troubled townspeople. In 1882, the *Cootamundra Herald* anguished over the 'wretchedness and brutalising' of Aborigines in an 'unprotected camp' outside town. How could settlers, the newspaper asked, who have prospered on the ruin of our country's own race, whom we have thoughtlessly disregarded, outraged, and ill-treated, look on such a scene without a reproachful conscience?[10]

Categorisations of Aborigines as primitive and unworthy dispelled settler shame and cast as reasonable colonial processes of dispossession. Aborigines, colonists told themselves, deserved displacement. In June 1890, the *Cootamundra Herald* published an article by Carl Lumholtz, author of *Among Cannibals*, an account of his Australian travels. 'The aborigines of Australia are the lowest and most degraded human beings to be found on the face of the earth', Lumholtz claimed. Aborigines had no 'higher ideas of religion'. They used crude tools of stone, timber and bone, and led lives 'of idleness, of robbery, fight and cannibalism.'[11]

Days later, the paper printed a list of areas in the Cootamundra and Gundagai districts opened for selection by the Department of Lands. One block of land consisted 'of moderately timbered box forest, red clay soil (upon which the timber has been killed by ringbarking) and may be classed as agricultural land.'[12] Did settlers recall the derogatory images created by Lumholtz as they rode past the Cootamundra fringe camp to inspect farmland for selection?

Belief in the inevitability of change helped justify the actions of settlers. Time itself became a driving force. In 1894, a reporter visiting Cootamundra described a profound transformation.[13] Thirty years previously, the swampy place reverberated with the calls of turtles, insects, frogs and wild ducks. The 'evolution of time' had brought extraordinary changes, the reporter wrote. No longer did people hear a cacophony of swampland species. As decades passed, the 'splendid town' of Cootamundra had 'sprung up'. The town site, the journalist mentioned, was formerly 'a lake or swamp', and watery areas were 'drained by the growth of the town'. Settlers were accorded little agency in the newspaper story of dramatic ecological change. Almost naturally, as time passed, the town grew and species vanished.

In fact, rather than experiencing an inevitable unfolding of history, settlers deliberately made social and ecological changes. Rumours swept Temora in 1890 that Reverend John Gribble planned to set up an Aboriginal reservation in the district. The newly established Progress Committee of Temora agreed to 'take the matter up and protest against such an infliction on their fair and rising district.'[14] The social and ecological realities known today on the southwest slopes were not delivered by an inescapable process of change. As the formation and deliberations of the Temora Progress Committee indicate, settlers actively constructed notions of progress to enable agricultural, commercial and social gain.

In the first decades of town development on the western slopes, civic leaders strove to construct streetscapes signifying progress towards European ideals of civilisation and comfort. Colonists, it appears, were anxious to convince themselves and visitors that imported systems of farming and British patterns of settlement would indeed prove stable and prosperous. 'Temora looked as if it might be much older than it is', observed a visiting journalist in 1909. The reporter noted

> neat cottages and pretty villas, embowered in many instances in a profusion of creepers and shrubs, and walking through the main thoroughfares of the town, saw fine substantial buildings lining the streets on either sides.

There were

> beautiful churches, schools, and other institutions, with the undulating country all round it cleared for some distance, and studded with comfortable residences.[15]

In 1900, a journalist found Cootamundra 'one of the cleanest, best laid out, and well-regulated towns in New South Wales.'[16] The mayor and aldermen had achieved 'remarkable work' developing the settlement. The journalist wrote:

> Its well-made, rectangular streets, planted with shade trees, its public reserves, its abundant water supply, its sanitary and lighting

systems, its many valuable buildings, pretty churches, hospital, municipal buildings, banks, school of arts, free public library, all tend to make a healthy and progressive town.

English traveller Michael Davitt visited southeast Australia in 1898. In the coastal cities he encountered 'wealth, progress, and enlightenment', then journeyed inland through rural areas 'which has made them what they are.' Davitt thought Wagga

> one of the prettiest little inland towns in the colony ... The streets of the town are wide and are planted with the pepper tree on each side. The houses are well built, and there are many handsome public buildings and churches. The countryside is dotted with choice residences and gardens, while vineyards and orchards are plentiful along the river banks.[17]

In 1938, the people of Wagga joined national celebrations marking 150 years of European settlement in Australia. The local council and chamber of commerce issued a commemorative booklet, *Wagga Wagga: a far cry*. The front cover showed an Aboriginal man and stylised modern buildings. According to the Wagga artist responsible, the design represented 'primitive man and modern civilisation.' The regional centre and Murrumbidgee River district had moved far down the linear pathway of progress. 'Could we but look with primitive wonder at modern Wagga and her beautiful countryside', the artist wrote. Anxiety remained in 1938, it seems, about the validity and permanence of British settlement on the southwest slopes. The sesquicentenary gave opportunities to construct and maintain strident stories of local and national progress. Perhaps an effusive advertisement published inside the Wagga souvenir booklet helped allay concerns:

> One Hundred & Fifty Years!!! Years of Glorious Endeavour & Achievement! Years that have seen our Vast, Beautiful Continent grow from a Primitive Wilderness into a Land of Plenty – a Land of Progress! A Glorious Outpost of a Glorious Empire! Gold! Wheat! Wool! Factories! Industries!

Agricultural science and industrial technologies transformed the western slopes in the twentieth century. Town development symbolised regional progress, a break from the past. In the 1960s and 1970s, growing public interest in Australian and local histories brought new ways of seeing places and old buildings. Identification of 'heritage' signified a separation between past and present.

The short history of European settlement in Australia made it difficult to imagine the past as past. As historian Tom Griffiths observes, historical events remained close, 'often astonishingly so', to the lives and times of settler Australians.[18] The identification and preservation of old buildings and sites of early settler activity as 'heritage' helped maintain the forceful idea of progress. Perceptions of historical divorce served those who looked to the future.

The year 1961 marked a century since the New South Wales government decided to establish the settlement of Cootamundra. An official souvenir of the centenary describes great strides taken by the town and district.[19] According to one contributor:

> A survey of the progress of agriculture in the Cootamundra district would cover such a wide field that in an article of this nature only significant phases can be mentioned.

Inside the souvenir publication, a reproduction of an old photograph captioned 'Parker Street in 1870' shows eucalypts and a few small buildings beside a dusty road. A second image sits below: '100 years of progress. A night view of Parker Street in 1961', the caption reads. Double-storied shopfronts and hotels line the main street, stripped of deep verandahs and bathed in electric light. Introducing the souvenir, Mayor Twomey honoured the pioneers of Cootamundra. Locals had an 'inescapable duty and responsibility', wrote Twomey, 'to carry on the work of progress, begun 100 years ago.'

Eight years later, almost two thousand Cootamundra town and district residents signed a petition opposing Cootamundra

Municipal Council plans to demolish the town hall office building and tower.[20] Cootamundra people felt 'a high regard and much affection' for the tall and decorative Victorian structure, claimed the petition organisers, and council had 'not fully appreciated the historical and architectural value of the building'.

A new administrative block and improvements to the dilapidated hall and kitchen would serve future generations admirably, aldermen explained in reply. 'We're looking 30 years ahead', Alderman French told the *Cootamundra Herald*.[21] When the replacement complex opened in 1971, dignitaries spoke of the inevitability and promises of progress. 'The completion of this major project is not only another step forward for Cootamundra, but we should also recognise it as symbolic of our progress', declared Alderman Richards, Mayor of Cootamundra.[22]

At the official opening, state politician John Waddy acknowledged those who had tried to save the original building from demolition. Unfortunately, the relentless push of linear time made its destruction inevitable:

> There is no one among us today who doesn't feel sad to see historical old structures removed when they are in the path of progress.[23]

Like the removal of old verandahs and iron lacework from main street hotels and shops, the construction of a new municipal office building represented movement towards the future. John Waddy told the crowd at the opening:

> A town's history is not only served by preserving its buildings of yesterday, it is also served by raising new edifices today that will stand for the benefit of the generations of tomorrow and the day after.

Clear demarcation between past and present enables economic activity and the dramatic transformation of places. Linear time is imagined as a simple imperative divorced from contingencies of place and history. Present consequences of past events hold no sway, as the past is firmly considered past.

According to social theorist Dipesh Chakrabarty, efforts to objectify the past reflect a desire to be free of the past.[24] Freedom from past events and from unfortunate outcomes of historical processes enables new realities to emerge. Buildings and places classed as settler heritage guide movement towards a brighter future.

In 2004 the Cootamundra Shire Council erected new road signs welcoming travellers to the district. A golden sun rises above productive green hills. 'New Country Living', the signs read. According to the Cootamundra Development Corporation, an 'enviable combination of then and now, of tradition and development … makes Cootamundra unique.'[25] The Cootamundra Shire Council website associates modern agribusiness with settler heritage:

> Cootamundra and district have always produced good beef, lamb, wool and rich crops of grain. A recent study has identified that Cootamundra has exceptional natural resources for Agribusiness Development.
>
> Cootamundra has always maintained an affinity with days gone by. Many of the older buildings remain, and are being sensitively restored, while in Cootamundra's Cooper St a whole avenue of century-old elm trees have been heritage listed.[26]

In a promotional booklet, Temora Shire Council likewise binds prospective commercial development to historical narratives of linear, progressive time. Temora has many heritage buildings, the booklet explains. 'Edwardian and Federation periods are well represented throughout the town'. A photo of a decorative old church overlaps another showing a wide paddock of flowering canola, a brilliant yellow display of successful modern farming. Looking forward to productive futures, the publication evokes the imaginary power of a separated past. Distinctions between past and present are intertwined with references to economic activity and agricultural development:

> The district is an agricultural showcase with grain and livestock dominating the landscape to the far horizons. Grain terminals

> bursting with activity at each harvest point to a rosy future. The diversity and strength of agriculture is the foundation of the local economy. The rural fabric of Temora makes it a good place to get 'grounded', to go back to our origins, to get back to basics and a simpler way of life away from the hustle and bustle.

Nostalgic constructions of the past allow ethical responsibilities to address needs arising from histories of ecological disorder and social injustice to be evaded. The past is imagined as a comfortable domain, separate from the present but clearly visible. 'The Riverina I live in has charm because it hasn't lost touch with its past', remarked a Riverina resident in a pamphlet promoting the region. History, the pamphlet suggests, is encountered inside regional museums and across the painted facades of heritage buildings, not in the everyday patterns of life. Dominant historical narratives obscure connections between brutal events of the past and realities of the present. Past events are seen to belong in the past, imposing no limits on human activities today.

In his 1968 play *The cherry pickers*, Wiradjuri writer and visual artist Kevin Gilbert challenged historical narratives blind to past injustices and the ethical responsibilities of settlers. Bungaree, a character representing a famous Aborigine of early Sydney, describes 'a web of lies to strangle human right'.[27] The 'burning native souls' of his people are 'aglow aglow demanding justice done'. Gilbert's character warns audiences to reject untruthful, nostalgic stories of Australian settlement:

> Look back and look beyond to history
> to know your country's birth, to know the truth
> or is your love so base — built on deceit —
> Then lo! The monstrous march of burning feet...

On the southwest slopes, old buildings freshly painted in heritage colour schemes today serve functions similar to those of the 'wilderness' imagined by earlier generations of progressively minded colonists.

In 1930 Temora residents celebrated fifty years of town settlement. An author of a souvenir publication interviewed William Marshall, one of the first settlers to arrive in the district. 'In the early fifties the solitudes of the big stretch of country lying between Cootamundra and Wyalong were almost unbroken', the elderly man remembered. A dark and lonely silence prevailed:

> No sound was heard except the 'neighing of the wild horses and the howl of the dingo.' It was in many respects, said Mr. Marshall, a wilderness. But the country appeared to be suitable for settlement, and it was being gradually taken up.[28]

Another long-time Temora resident spoke of Wiradjuri spearing cattle and killing sheep, 'but as civilisation advanced the complaints about their conduct became less and less.' Wild scrub and Wiradjuri people vanished as brave pioneers responded to the imperatives of progress, recounted the jubilee souvenir:

> It mattered not that the landscape was wreathed in almost impenetrable forest and held by savage hordes, the pioneers' watchword was 'Advance', and the crack of the stockwhip was followed by the ringing sound of the bushman's axe. It was but another exemplification of the dictum that the fittest survive. The unenterprising, non-progressive native was dispossessed and driven back and the zone of civilisation was gradually extended.

The *Temora Independent* reprinted the souvenir writings of 1930 alongside fresh material in a 1980 commemorative edition marking the centenary of Temora.[29] Like the earlier publication, the centennial edition fifty years later presented bright, progressive narratives of local history, honouring those men 'who made the great heart' of the Temora district 'yield to their labours'. Late in the twentieth century, people remained faithful to ideals of progress. 'May the progress and achievements of our first 100 years be but a starting point in a never-ending march to greater achievements', the centenary supplement declared.

Ron Maslin helped found the Temora Rural Museum in the early 1970s. Inside an enormous iron shed, Ron and I examined antique grain picklers, a hay rake, horse drawn harvesters and other old agricultural machines. Ron carried a booklet inscribed with the dates each item was invented. He named Australian inventors of revolutionary farming technologies – Richard Bowyer Smith, HV McKay, Headlie Shipard Taylor and others. I asked Ron to explain the role of the Temora Rural Museum. The museum preserved outdated machinery and artefacts, he replied, to convey a story about 'the progression' of farming technology from the early days to the present. Curators worked 'to preserve the history' of rural life.

The last piece of machinery we looked at was a stump-jump plough made in Temora in about 1900. Richard Bowyer Smith, a South Australian farmer, invented the stump-jump plough in 1876, Ron explained. Farmers found imported ploughs unsuited to Australian conditions. Farmland cleared hastily of trees retained stumps and roots that snagged regular, fixed-blade ploughs. Buried stones caused problems too. Drivers and horses were often injured, and ploughs damaged. Levers on the implement invented by Smith allowed plough blades to ride over buried roots and stones. 'The only time you might get trouble', Ron said, 'is if they get under a stump where there's a big root, and they jam.'

I picked up a pamphlet about the Temora Rural Museum as I left the building. 'Preserving the Past for Tomorrow's Generation!' the pamphlet declares beneath a photograph of a restored steam engine.

On the main street, a newsposter in a mesh frame carried the latest *Temora Independent* headline: 'Commercial release of GM canola proposed'. I bought a copy. Two transnational corporations had applied to the federal government for approval to sell genetically modified canola seed to Australian farmers. Another step along the inescapable pathway of progress? Yet again, constructed notions of linear time and modern imperatives disguised and naturalised choices made by powerful individuals and groups to advance their own interests.

'You can't go back'

In 1967 Colin Donald, Professor of Agriculture at the University of Adelaide, published 'The progress of Australian agriculture and the role of pastures in environmental change'.[30] In the academic paper, the eminent agricultural scientist presented a triumphal narrative of scientific progress. Agricultural scientists, Donald noted, first made a substantial contribution to Australian farming late in the nineteenth century. Wheat yields were falling as cropping depleted soils of essential nutrients like phosphorus and nitrogen. Applications of superphosphate fertiliser, scientists found, reversed the decline.

Other problems appeared in the first half of the twentieth century. Soil structures deteriorated as farmers repeatedly cultivated paddocks to control weeds and conserve moisture. Gullies fragmented hillsides, and windstorms carried topsoil from paddocks. Agricultural and soil scientists, Donald observed, solved erosion problems too. In the 1950s and later decades, leguminous pastures and chemical farming systems restored the fertility, organic content and erosion resistance of soils, and boosted agricultural production.

Donald's paper contains a number of graphs. Diagonal lines chart steady rises between World War II and the late 1960s in sheep numbers, wool production, average fleece weight, and the total area sown to 'improved' pastures.

The linear, progressive narrative articulated by Donald continues to drive the work of agricultural scientists. At the start of the twenty-first century, scientists hope to solve escalating problems of soil acidification and salinisation, the latest in a line of challenges facing Australian agricultural production.

Abstract notions of linear time and industrial space obscure the dynamic nature of rural places, enabling subjugation and intense agricultural activity. When the needs of local ecologies are denied to enable movement towards 'the future', unexpected obstacles inevitably arise. Modern industrial agriculture, observes scientist and philosopher Vandana Shiva, is 'based on the assumption that technology is a superior substitute for nature, and hence a means of producing limitless growth, unconstrained by nature's limits', by the particular realities of the land. Perceptions of nature 'as a source of scarcity', she continues,

> and technology as a source of abundance, leads to the creation of technologies which create new scarcities in nature through ecological destruction.[31]

Calls for attention towards the ecological disordering of farmland disturb the dominant narrative of progress informing agricultural science and industrial, export-oriented agriculture. Representatives of modern farming defend and reinforce linear historical perspectives by accusing critics of wishing to erase all traces of colonisation. In October 2002, *The Land* newspaper attacked those 'green groups and environmentalists who think the only way forward is to go backward'.[32] Agriculture had progressed far along the development path, the newspaper insisted. 'Trying to back pedal to 1789 isn't an option.'

Likewise, agricultural scientist David Smith writes of 'those people who long to "go back", to recreate the past.'[33] Such aims can never be realised, Smith argues. Rather than 'fruitlessly attempting to retain what is, or fruitlessly working towards what is believed to have been', people should instead focus on the great responsibility of 'future-making'.

Directing attention towards the future, and devaluing 'what is' – the particular natural patterns of rural places – serves modern, commercial interests. Smith imagines a linear concept of time abstracted from the dynamic and expressive life of places. Such a notion builds perceptions of inert and malleable space, a 'natural resource' available for industrial agriculture and future commercial activity. Throughout the history of Australian colonisation, linear concepts of time and the imaginative replacement of lively places with malleable spaces enabled settlers to harness land for the British Empire, the Australian nation, the global economy and themselves.

In *The cherry pickers*, Kevin Gilbert offers an understanding of time alternative to the progressive, linear concepts dominating settler Australian culture and modern farming. Generations of Wiradjuri families have travelled to the Young district as summer begins to harvest cherries ripening on hillsides. In the 1950s, families came to the Young cherry harvest from Brungle, Cowra and Condobolin, then travelled to Griffith to harvest apricots.[34] The work was hard and poorly paid. *The cherry pickers* explores issues of culture, history and justice as a group of Aborigines join the annual cherry harvest.

Towards the end of the play, Tommlo, a young man, argues about notions of progress and Aboriginal identity with his wife Zeena. Tommlo plans to reconstruct and perform a traditional ceremony. Zeena opposes the idea. Tommlo decorates his body under an old cherry tree. It is summertime, and the weather is especially dry and hot. Wind carries dust across the stage, conveying a troubling sense of disorder and instability. Zeena thinks it particularly absurd to hold the ceremony beneath the elderly cherry tree: 'Doesn't this prove that *some* advance has been made because "cherry tree" means money – *and* food?' Aborigines shouldn't embrace 'a stone-age identity in a nuclear age', she insists. 'We must advance, must mature and must never revert back', Zeena declares, 'for life is a constant process of growth.'

But Tommlo holds a different, less linear concept of time. He imagines time bound to place and natural cycles, not abstracted from land. Tommlo realises he 'was lost' before, when he 'looked at life, the world, the whiteman's way'. He rejects abstract concepts of linear time and delights in a personal identity embedded within a dynamic web of life:

> I ain't lost anymore. I am a *nothing*. The trees, the grass, the river, the earth is life, is *everything*. I am *nothing*, a *nothing*. Now that tree is *me*. It is all of me. I am that tree. I am nothing, yet I am

somethin' because the earth is me. These rocks are me and I am the movin' soul of them all. See, I looked at the tree and said that is a tree. I kept it all separate and alien, but now, like the old days, I am a nothing but that *tree* is me and I am a something and when I die I will flow into the creative essence that made *me*, the tree and all created life, for we are all inseparable.

Tommlo convinces Zeena to join the ceremony and dance. Later, characters notice changes induced by the performance. Bubba, an elderly woman, discovers a renewed feeling of place, of land in lively motion. She is shocked to see nature faltering:

> It's like as if the Old Days have come back again. Everythin' is movin'. Even the ground is sort of livin' again – or is it – is it dyin'???

Oral, indigenous peoples, philosopher David Abram observes, don't share the western 'distinction between a linear, progressive time and a homogenous featureless space'.[35] Instead of abstract concepts, time blends with the physicality of places, enlivening rich and dynamic terrains. Not divorced and linear, time is embedded and active. Land is shifting, vital. In *The cherry pickers*, Bubba begins to sense the wounds of local nature when she embraces again a Wiradjuri framework of perception and understanding. 'Only when space and time are reconciled into a single, unified field of phenomena does the encompassing earth become evident, once again, in all its power and depth', Abram writes. Rejection of an imaginary western division between time and space enables both Bubba and Tommlo to sense the lively motion of trees, rocks and earth, the dynamic patterns of place.

Efforts to modify ecologically disruptive practices are often informed by the same progressive notions underlying the disordering of natural systems. Aiming towards idealised future goals like 'sustainability' tends to obscure the immediate needs of places, people and other species.

At Gundibindyal, east of Temora, the Grain Research and

Development Corporation and other agribusiness organisations trial new farming methods and crop varieties. 'AUSTRALIA'S GROWING FUTURE, HIGH PERFORMANCE CROPS – SUSTAINABLE FARMING', a sign reads beside the road. Farmland, the sign suggests, will some day endure the intense demands of industry and global commerce. There is no need for radical or immediate change, the wording implies, nor should we ask if too much is being demanded of Australian farmland today. 'By projecting the solution somewhere outside the perceivable present', writes Abram, every utopian vision of the future dangerously 'invites our attention away from the sensuous surroundings, induces us to dull our senses, yet again, on behalf of a mental ideal.'[36]

An embedded concept of time grants motion and dynamism to rural terrain. Possibilities appear for immediate awareness and actions fostering ecological connectivity and wellbeing. Lively places become visible where once only malleable spaces extended. Expressions of other species and natural forces present in farmland arise. The dynamic potential of ecological connectivity becomes apparent. Dialogue and regeneration begins.

Returning to Cootamundra

At the National Library in Canberra I found a photo taken in 1922 from Billygoat Hill, a stony rise on the southern edge of Cootamundra. The growing town extends across the top of the image, beneath a curving horizon studded with eucalypts. Amid street trees and gardens are houses, public buildings, a church steeple. Granite outcrops protrude from the grassy slope where the photographer stood. A line of slab fence posts leads down Billygoat Hill towards swampland. Pools of water reflect mature eucalypts. The swamp surface is uneven and textured. Dark clumps of taller plants emerge from the grassy expanse.

Gazing at the image, I imagined frogs calling and a yellow-billed spoonbill wading.

'Aborigines never lived here, away from the rivers', some people tell me. 'Sure there are stone tools in the paddock', they might say, 'but the people who made and used them were just passing through, journeying between the Murrumbidgee and the Lachlan or eastward to the coast.'

The common stereotype is a simple binary of settled colonists and nomadic, wandering tribes – a useful notion, granting validity to colonial processes of dispossession by refuting Aboriginal connections to particular places. Denying the humanity and dignity of Aboriginal people also enabled colonisation. Settlers imagined themselves 'civilised' and 'modern'. They classed Wiradjuri as 'savage' and 'primitive', the black and homogenous 'Other' to European whiteness, soon to vanish as agricultural development progressed. Such denials and representations allowed colonial activities to proceed unhindered by ethical constraints.

When the photograph of the swampy area Cootamundra residents called 'The Flat' and 'Frogs' Hollow' was taken in 1922, there probably remained Gudhamangdhuray clanspeople who remembered tending the watery place as a freshwater turtle sanctuary, before the arrival of the Great Southern Railway and the growth of the town. Growing up in the district in the 1970s and 1980s, I heard little about the rich origins of the placename 'Cootamundra', nor stories about the erasure of Gudhamangdhuray swampland and people. Colonial power, literary critic Edward Said observes, creates 'the security of a situation that permits the conqueror not to look into the truth of the violence he does.'[37] At Cootamundra, silence obscured detailed meanings and narratives, blocking possibilities of ecological and social justice.

Today, brick veneer homes and tidy suburban gardens cover the northern slope of Billygoat Hill. One summer afternoon I walked from the top of Poole Street over the western side of the granite rise and down towards Cootamundry Creek. A tangle of kangaroo and wallaby grass, red gum saplings, spreading flax lily, hopbushes and St John's wort cloaks the hillside behind the golf course. Hurleyville homestead, a red brick house built by the Hurley family, squatters of Cootamundra station, stands to the east, on the southern slope of the hill. A small freshwater spring in the bed of Cootamundry Creek, an underground source never known to run dry, perhaps led John Hurley to build the headquarters of Cootamundra station here.[38] Gudhamangdhuray people living in a station camp near the homestead probably considered the spring of much significance. Settlers called the watery place 'Hurley's Springs'.

Cootamundra prospered when linked to the port city of Sydney by the Great Southern Railway in the late 1870s. Residents reluctantly drew water from a polluted reservoir on Muttama Creek, near the centre of town. In 1881, the *Cootamundra Herald* announced an 'Important Discovery'.[39] A government engineer had sunk a well at Hurley's Springs and 'tapped a fine strong flow of pure crystal spring water'. The engineer declared the source abundant. The editor of the *Herald* imagined 'playing fountains in our Park; and every householder a fine shower-bath in his upper storey'. Townspeople could at least hope for 'something better than the stagnant corruption of the reservoir, that almost causes their cattle to vomit.'[40] In time, water drawn by steam engine from Hurley's Springs into a storage tank on Billygoat Hill gushed into Cootamundra homes.

As the town grew in the first decades of the twentieth century, residents looked beyond Hurley's Springs for more secure supplies. Inside the town hall early in 1933, jugs of water piped from the Murrumbidgee River lined tables set for dinner. Guests filled tumblers and rose to toast 'The advent of the river water supply'.[41] Beside the Murrumbidgee at Jugiong, great pumps pushed water eighteen kilometres uphill to a gap on Cowang Ridge, halfway to Cootamundra. Gravity completed the delivery.

Today at Hurley's Springs, a large dam holds water where the slope meets creek flats. In the afternoon heat of summer, I walked onto a clay and gravel surface. The reservoir had receded since winter – evaporated, drunk by kangaroos and pumped over the rise to irrigate the golf course. Lying about were many shells of freshwater mussels, a food species gathered by Wiradjuri.

Alongside the shell remains grew flowering plants, perhaps old man weed. I remembered Bob Glanville's stories about the herb with a pungent aroma, a renowned medicine plant. His family, members of the local Gudhamangdhuray clan, harvested old man weed from a dam on Cowcumbla Street, on the other side of Billygoat Hill. Bob hadn't seen the plant growing in the Cootamundra district for decades. Chemical farming destroyed it, he suspects. Powerful medicine was made from old man weed, Bob told me:

> You'd boil it up. If you had stomach-ache you'd drink the juice. If you had respiratory problems, if you had a bad cold, if you had flu or something, you'd put a towel over your head and get the vapours out of it. Or, if you had a wound, or a sore, or a rash, you'd bath it in the juice of the old man weed. Oh it worked, it was great.

I rang Bob a few days later. We met at the top of Poole Street and walked over the hill, down towards Hurley's Springs. The plant growing beside the dam, Bob confirmed, was indeed old man weed. Beside the dam we crushed leaves and stems between our fingers and whiffed the spicy scent.

As we marvelled over the medicine plant, Bob told me about the time he went yabby catching with his siblings and cousins in flooded swampland near their home on Cowcumbla Street, then a rough track running along the northern edge of Billygoat Hill. They used kerosene tins to dam a drain gouged across the grassy paddock. Bob cut his shin on rusted tin and the wound became badly infected. When a local doctor couldn't contain the infection, his grandmother, Melinda Bell, took charge. She collected old man weed, chopped and boiled the green leaves and stems, then treated Bob's leg with the solution. The doctor was amazed at how quickly the wound healed.

We walked back up the slope towards town. Recently, at the Riverina Archives in Wagga, I'd found an old story by James Gormly about Cootamundra station, and shared with Bob what details I could remember.[42] In the 1840s, Gormly lived at Nangus, near the Murrumbidgee River west of Gundagai. There he knew an elderly and respected Aboriginal warrior who settlers called Billy the Ram. Many years later, a surveyor described to Gormly the burial of Billy, an event he'd witnessed in 1859. The old man died in 'the blackfellows camp' on Cootamundra station. Riding past, the surveyor saw Aborigines digging a grave near Cootamundry Creek. He noticed Billy's body wrapped in a blanket and a possum skin rug. Bob wondered if the grave was near where an old mud hut once stood on the Junee road. His grandmother Melinda always warned the children to stay away from the decrepit structure. She told them an evil ghost, a 'bageeyn', lived there.

At the crest of the hill we passed a stack of broad steel pipes. The pipes were to be buried across Billygoat Hill, Bob explained, as part of a new effluent recycling scheme. Several dams, including the Hurley's Springs reservoir, would hold treated sewerage water. The scheme would halve the volume of treated effluent, dangerously rich in salts and nutrients, sent from the Cootamundra sewerage plant down Muttama Creek. The shire council planned to irrigate sports ovals, town parks and the golf course with the recycled water.

I wondered whether the freshwater mussels in the Hurley's Springs dam could live in salty, treated effluent. And would old man weed still grow? Once the pipes were laid, the water level of the dam would remain high, even as summer approached. No longer would falling water offer the bare, moist expanses the medicine plant needed to germinate and grow.

Driving down Poole Street, we passed a wide concrete depression skirting a metal grate. A buried network of stormwater drainpipes ensures that the low-lying terrain between Billygoat Hill and

Muttama Creek rarely floods. Where Bob and his family once caught yabbies and ducks there are homes, a primary school and soccer fields.

As a young man in the 1960s, Bob had watched workers grade drains, bury wide concrete pipes and form streets across Gudhamangdhuray swampland. Housing Commission contractors erected fibro and brick veneer houses.

Bob couldn't remember talking to his grandmother about the destruction of swampland at Cootamundra. In 1971, the year Melinda fell sick and died, workers had built a football oval near the Mercy Hospital across a watery swampland remnant.[43] Bob imagined that the erasure of remaining swampland south of Muttama Creek saddened his grandmother. 'That part of the town was a very significant place for her', he explained, 'a very significant place.'

An oil painting hangs in the entrance hall of Bob and Vonnie Glanville's Cootamundra home. The picture shows trees, swampland and streetscapes – the terrain captured by the anonymous photographer from Billygoat Hill in 1922. Bob had a local artist render the work from a different photo taken years later, probably in the 1930s, from a position further east on Billygoat Hill. In the background lies the town – houses, street trees, the Catholic church. Closer to Billygoat Hill, fences divide swampland into paddocks. A cottage sits in the foreground, where the slope begins to climb. Facing an unformed dirt road – Cowcumbla Street – the corrugated iron home has a small front verandah. A clipped hedge and timber fence define the front boundary. Distanced from Cootamundra by a wide stretch of swampland, the house belonged to Melinda Bell, and was Bob's childhood home.

Marie McGuiness gave birth to Melinda in 1896 at Brungle Aboriginal Station, a government reserve near Tumut, east of the Murrumbidgee River. The reserve population ebbed and flowed throughout the late nineteenth century. Wiradjuri made regular journeys between government reserves – at Brungle, Cowra, Yass and Warangesda - and unofficial fringe camps, visiting family and maintaining ties to country. Sometimes over a hundred Aboriginal people, mostly Wiradjuri, lived on Brungle reserve. Other families camped nearby, beside the Tumut River.[44]

Wiradjuri came from far away to live at Brungle. Townspeople didn't welcome Wiradjuri families and individuals displaced by agricultural development and closer settlement. Late in the nineteenth century, town councils at Gundagai, Tumut, Yass, Cowra and Cootamundra demanded that the Aborigines Protection Board disperse town fringe camps.[45] Like Warangesda down the Murrumbidgee and Hollywood mission at Yass, Brungle Aboriginal Station was located in pastoral country, outside the more productive and intensively managed agricultural districts. Government reserves offered food rations, accommodation and the promise of contact with kin.

At Brungle, Wiradjuri blended traditional ways with settler customs. Until the 1920s, boys underwent initiation rites on Mudjarn, a mountain overlooking the reserve. Vince Bulger grew up on Brungle reserve in the 1930s and 1940s. He remembers initiation scars across the chests of old men.[46] A photograph taken at Brungle in 1894 and held by the Australian Museum in Sydney shows men holding boomerangs and spears. Partly clad bodies are painted with stripes, dots and European letters.

Arthur McGuiness, Melinda Bell's father, was a Gudhamangdhuray clansman. He worked as a horse breaker on stations in the Yass and Gundagai districts. Arthur may have learned his skills on Cootamundra squatting run, where he probably grew up. The station was known for its great herd of feral horses. In the 1840s, James Gormly helped muster the wild descendents of thoroughbreds sent by absentee Cootamundra squatter John Hurley from his property near Sydney. Breakers had a hard job training the frenzied animals.[47]

Melinda's elder brother, named Arthur after their father, was born at Cootamundra. Bob Glanville thought his grandmother probably spent time there as a child. Families moved around, he explained. They travelled from Brungle up Muttama Creek to Cootamundra, then sometimes northeast to Cowra. As an old man in the 1960s, Melinda's brother Arthur told ethnologist Janet Mathews that Cootamundra was his 'ngurambang', his 'native home'.[48]

In July 1916, the New South Wales parliament granted the Aborigines Protection Board powers to remove children without approval from parents or the judiciary. The board instructed managers of Aboriginal reserves across New South Wales to identify all fair-skinned and neglected children.[49] Over the first two months, the board took eleven kids from Brungle families. Girls went to the Aboriginal Girls Training Home at Cootamundra, boys to a similar institution at Bomaderry, north of Sydney.

Margaret Tucker lived at Brungle as a child during World War I. She remembered the scarcity of other children: 'Most had been taken away to be trained, never to be seen for many years.'[50]

Melinda McGuiness feared losing her young, fair-skinned daughter Iris, Bob Glanville's mother. In about 1918, she fled with Iris from Brungle to Cootamundra, the heartland of her father's clan. Her decision was probably fraught. Cootamundra was dangerous terrain for Aborigines. According to historian Peter Kabaila, the town's fringe camp dispersed when the Cootamundra Aboriginal Girls Training Home opened in 1912.[51] Wiradjuri distanced their children from the new institution and the close reach of Aborigines Protection Board officers. Cootamundra, the place itself, must have called powerfully to Melinda, drawing her back.

Melinda and Iris camped at Hurley's Springs, near the old headquarters of Cootamundra squatting run, beside Cootamundry Creek. Perhaps she returned to the old Cootamundra station campsite, where as a child she may have visited Gudhamangdhuray kin with her parents.

Melinda and her daughter later joined a poor community on

the western fringe of town, where people lived in self-made shacks. Melinda worked as a nanny for a prominent Cootamundra family. Bob imagined that the job lent his grandmother some status, helping to deflect the gaze of Protection Board officers. She saved enough money to build a small corrugated iron house beside swampland on Cowcumbla Street, then a rough track skirting the lower slopes of Billygoat Hill, devoid of other homes.

Melinda's small house soon filled with people. She married John Bell, and the couple had two sons. After John died, Melinda married his brother Bob, with whom she had a daughter. Melinda's niece, Minnie McGuiness, also came to live in the Cowcumbla Street cottage. Arthur McGuiness, Minnie's father and Melinda's elder brother, had lost his wife and thought the Aborigines Protection Board would take Minnie. Melinda travelled to Gooloogong on the Lachlan River to bring Minnie home.

Iris, Melinda's eldest child, married Bill Glanville in 1938. They too lived in the cottage on Cowcumbla Street. Iris and Bill Glanville had five children. Bob was their first child. He remembers sharing a double bed with four or five other kids. They slept under 'Wagga rugs' – chaff bags sewn together and quilted with cloth – 'and we were as warm as toast during those cold Cootamundra nights', Bob recalled. Journeying kinsfolk sometimes arrived on night trains, and Bob woke to find strange children asleep beside him.

With many children in the small house, Melinda feared the attention of Aborigines Protection Board officers. She strove to maintain a respectable position in the town. When her job as a nanny came to an end, she found work in homes and as a cleaner at the primary school. 'She was a nice old lady, old Mrs Bell', long-time Cootamundra resident Alec Hansen told me. Melinda worked for his aunt:

> She ironed beautifully and washed and sewed, and did all sorts of

things. She was a very skilled person, at those things that she'd become familiar with. And there was no reason that if she had been involved in some other trade or whatever, that she wouldn't have been skilled in that: a very intelligent woman.

Several years after she left Brungle, Melinda's brother Eric joined her at Cootamundra, ngurambang of the Gudhamangdhuray clan. Melinda was settled and happy, and Eric hoped his family might do well too. He and his wife Martha built an almost identical cottage next to the Bell family home on Cowcumbla Street. Eric found work as a horse breaker and station hand.

Bob remembers Melinda sitting with Eric on her front verandah. Alongside his cousins and siblings, Bob hid in the shrubs at the side of the house and listened to the elderly pair speak in Wiradjuri. Melinda and Eric would scold the children when discovered, and shoo them away. They didn't want the young ones to learn the language. Public displays of Aboriginality invited danger, Bob explained, and Melinda hoped her family wouldn't stand out:

> She didn't want us kids to overly identify, even though she knew we couldn't help being identified as being Aboriginal: the way we lived, like we were the fringe dwellers, and obviously, you know. She'd try to spare us a lot. We always had to be spotlessly clean. She was always spotlessly clean, in everything she did.

Rheumatism and arthritis slowed Melinda as she aged. 'She was only about seventy-five when she died', Bob said, but 'had worked hard all her life, you know, on her hands and knees scrubbing floors and all of that sort of thing.' In time, 'she just got old and tired.'

Melinda died in the winter of 1971 inside the Mercy Hospital, a red brick Catholic institution on the western side of town. Bob remembers vividly his grandmother breaking 'into a death chant' just before she passed away. Propped up in a hospital bed, Melinda sang in Wiradjuri with strength as her aged body failed. Her chanting seemed to echo throughout the hospital: 'it just went on, it was so repetitive, the same words, the whole time.' The moment was powerful. 'Didn't it make the hairs on the back of your neck stand up', said Bob. As Melinda chanted and death neared, Bob sensed the depth of her cultural knowledge,

> and I thought you know, oh you knew all that and you never ever taught us. You know she had all of that culture and that's all going to die with you.

The resonant chanting alarmed and confused the Catholic nursing sisters. 'They were absolutely in a state of shock. They were just scurrying around, going nowhere, but just scurrying everywhere'. Dressed in flowing white habits, several nuns knelt in the corridor outside Melinda's room and began to pray.

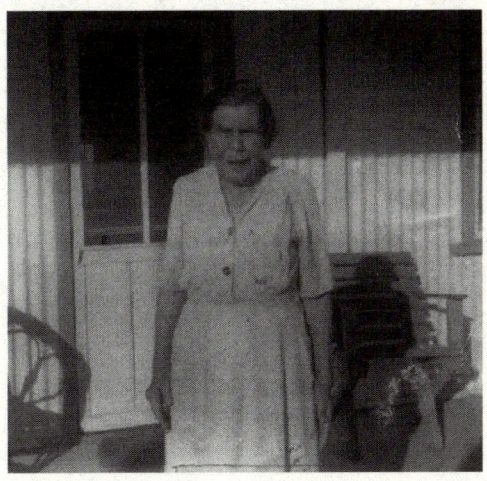

Melinda Bell on the verandah of her Cowcumbla Street home.

Melinda was buried in the Cootamundra cemetery, where her father, Arthur McGuiness, also lay. Bob reflected on the way colonial power affected the behaviour and outlook of his grandmother. The danger of passing on cultural knowledge had made her angry and sad, he thought. Melinda always refused to answer his questions about history and places. 'Oh no no, don't want to talk about that. Too sad son', Melinda would say.

Despite her caution and reticence, the Wiradjuri clanswoman did, in many ways, offer Bob and his family distinct Gudhamangdhuray identities. The family was descended from 'the old Cootamundra clan which lived here', Bob learned from Melinda and from his mother. Melinda taught her children and grandchildren 'the family whistle', Bob explained, a particular call of the pied currawong, the family totem. Melodious calls of currawongs fill Cootamundra streets during winter, when the large black and white birds descend from cold mountain forests to find food and warmth on the western slopes and southern tablelands.

Bob told me of a recent visit to the horse races in Canberra, where he spotted his brother-in-law in the crowd from the racecourse grandstand. Bob hadn't seen him for over a year. 'I gave him the family whistle', Bob said, the currawong call taught to family members by his grandmother. 'He nearly done a back-flip trying to find me, he recognised it straight away.'

Despite her general reluctance to pass on the language, Melinda did teach the younger generations of her family a scattering of Wiradjuri words and phrases. A second language sometimes proved useful, Bob explained:

> Culturally, she only taught us just enough for her to communicate with us kids, when we were kids, you know, with the language and things like that. She taught us enough about the language so she could communicate with us up the street without anyone else knowing! And those words stuck to us, and the family only know a few words, a few sentences, a few phrases.

As a young man, Bob travelled about the region playing football and fighting in a boxing troupe. Melinda would always go too, unless the destination was Wagga. The riverside place, Bob learned from his grandmother, was dangerous terrain, 'bageeyn country, evil spirit country'. When Bob spent a day at Wagga, Melinda refused to sleep until she saw him arrive safely home. Melinda told the children stories of a bunyip stalking the Wagga area, an ominous creature with powers to change shape. Bob remembers sticking close by his mother as a child when the family shopped in Wagga.

Melinda smoked the cottage on Cowcumbla Street when a child fell sick. She warmed the blade of a garden shovel in the fire then walked through each room, kangaroo dung and eucalyptus leaves smouldering on red-hot metal. 'What are you doing Nan?' the kids would ask. 'Oh, make the house smell better, make the house smell better', Bob remembers Melinda replying, evasively. Later, Bob realised his grandmother smoked the house to dispel 'the boorik', the evil spirit making the children sick.

When someone caught a cold, the kids went out with Melinda to harvest eucalyptus leaves. They would climb a tree and begin picking. Bob's grandmother ensured they picked only certain clumps of leaves: 'No not that one! That bunch over there', he remembers Melinda saying. Melinda boiled the leaves to make a solution for drinking, or instructed the children to chew them.

Bob told me how, as a Gudhamangdhuray clansman, his lifeways today belie settler mythologies of Aborigines wandering the land without close relationships with particular places. Before travelling to other towns in the region to attend meetings and conferences, Bob asks local Wiradjuri representatives for permission to speak. He acknowledges the existence of different Wiradjuri groups, and recognises the rights of groups over particular sections of tribal land. These rights and responsibilities include the power to grant or bar access to places and resources, a defining feature of Wiradjuri, the 'no-having' people of the western slopes and plains.

Melinda Bell's bustling household on Cowcumbla Street sat beside what remained of the freshwater turtle sanctuary. Before and after colonisation, Gudhamangdhuray clanspeople identified with the swampland and its creatures. Melinda could have built a house on land purchased elsewhere on the fringes of Cootamundra, Bob explained. She chose a site beneath Billygoat Hill on Cowcumbla Street, an unformed road without electricity, beside what remained of the sacred Gudhamangdhuray swamp. The low-lying paddocks between the cottage and Cootamundra were often wet. Waterbirds and frogs called beside rushes and old eucalypts. Melinda and her family caught yabbies and redfin. Bob remembers seeing swans. When Muttama Creek overflowed after heavy rain, the children paddled canoes across the flooded paddocks and trapped wild ducks. Returning to her father's ngurambang, Melinda allowed the place of rounded granite hills and swampland to claim the hearts and minds of her family – 'my beloved Cootamundra', wrote Bob.[52]

Modern educational and economic institutions disable the emergence of deeper ethical relations between people and local ecologies, biologist and conservationist Aldo Leopold argued in the 1940s. Western cultural processes drive people away, he observed, from 'an intense consciousness of land.'[53] Narratives of progress block awareness of local places. Movement and change are honoured, relations of commitment and stability devalued. In rural Australia, orientation towards distant markets and future horizons obscures natural patterns and particularities alive and active now. Scant human consciousness of local ecologies imperils living systems. As philosopher Val Plumwood observes, power rushes into vacuums of disengagement.[54]

On the southwest slopes, eviction and suppression of Wiradjuri people enabled the development of export-oriented systems of primary production. At Cootamundra, within an intensely colonised agricultural district, Melinda Bell maintained family relations with Gudhamangdhuray heartland. It was a remarkable achievement.

division

Farmland and 'the environment'

On a hillside of shale and quartz, inside an official 'nature reserve' on the southwest slopes, the trunk of an elderly ironbark is black and furrowed. A fence divides the slope, beyond wattles, scribbly gums and more ironbarks. Yellow box trees stand in the paddock below, where sheep graze an 'improved' pasture sown into deeper, richer soil.

Trees express particular knowledge of local terrains. Ironbarks indicate shallow, poor soils, yellow box relatively deep, fertile earths. The fence line dividing farmland from nature reserve, yellow box from ironbark, reflects a defining characteristic of western agriculture: fragmentation of natural patterns. Making rural places 'yield the produce upon which agricultural life depends', anthropologist Hugh Brody explains, involves

> separating manipulable resources from the rest of the environment and working with determination and consistency against all that might undermine this endeavour.[1]

National Parks and Wildlife Service rangers manage a number of nature reserves on the southwest slopes. Bushland skirts The Rock, a monumental stone hill rising above farmland southwest of Wagga. The Rock Nature Reserve 'is an island of natural habitat for native animals, including the turquoise parrot and glossy black cockatoo', the National Parks and Wildlife Service website reads.

Like other nature reserves in the region, The Rock Nature Reserve is a stony place of relatively impoverished soils within a sea of farmland. Sharply dividing domains into 'nature' and 'culture' is a peculiarly western activity, a product of Enlightenment and more ancient philosophical traditions. In the Australian agricultural imagination, farmland is associated with culture, bushland and revegetation plantings with nature. Such categorisation reflects what philosopher Freya Mathews calls 'the compartmentalised mentality of modernity', a system of knowledge by which everything undergoes division and segregation.[2]

The logic of modern agriculture divorces primary production from natural systems. Farmers today rely on industrial technologies and manufactured inputs, not ecological relations and nutrient cycles, to produce food and natural fibre. Rarely are diverse local communities of trees, shrubs, insects, birds and other animals seen to play a central role in maintaining the productivity of rural places.

Denial of natural connectivity underlies the popular notion of 'crop protection'. Avcare represents Australian farm chemical and biotechnology companies. 'Crop protection products', the organisation explains, 'control diseases, insects, pest animals and weeds which harm or destroy our food and fibre crops.'[3] Crop monocultures are vulnerable, and farmers depend on poisons to destroy potentially devastating insect populations. According to Avcare, insects are only trouble for farmers and crops:

> Insects can significantly reduce crop yields and quality through their feeding. Insect damage also assists the entry of bacterial and fungal diseases thus further reducing the value of the crop. Insects such as aphids or scale insects can also carry virus diseases from plant to plant and insecticides help minimise this damage by controlling insect pests.

The rhetoric of crop protection casts ecological relations and biological diversity as problematic. Insects interfere with monocropping and must be destroyed. In the same manner, modern farmers seek tight control over weeds. Across open paddocks, herbicides eliminate species classed as competitors with crop plants for sunlight, moisture and nutrients. Industrial monocropping, where vast paddocks are cloaked in single crop species, 'is agriculture as an engineer might conceive it to be', biologist Rachel Carson observed in her seminal book *Silent spring*.[4]

Modern agriculture takes great risks to generate abundant harvests. When chemical sprays prove ineffective, or when droughts, storms, frosts, bushfires and other natural phenomena appear, severe crop failures occur.

Responding to the major drought of 2002, member for Murrumbidgee Adrian Piccoli conveyed his understanding of a firm distinction between nature and agriculture. Farmers were at war with Australian nature, the politician explained, an unpredictable and hostile entity:

> Drought is one of those almost intolerable hardships Australian farmers have had to endure since the first European settlers began to open up vast tracts of land in the 19th century. We have toughed it out through the dry times before and, together, we will do it again. Overcoming the difficulties brought about when nature turns against us is never easy. It is just as well those on the land have inherited the never-say-die, fighting spirit of their forebears. It is just as well Australian farmers are the best at what they do in the world. They have to be.[5]

In 1999 Environment Australia held a conference called 'Balancing conservation and production in grassy landscapes'. An inescapable trade-off exists, the title implies, between primary production and nature conservation. People are not part of nature, westerners tend to believe, and so farming can't be integrated within natural systems.

Dedee Woodside, an ecologist working in the Murrumbidgee Irrigation Area, recently articulated the popular view of farmland as distinctly cultural terrain, existing to nourish only humans. We must 'sacrifice' nature to keep ourselves alive, Woodside argued:

> So I guess what we're getting to a point of is understanding the landscape a whole lot better, to be able to say: 'this area is good for this function, and should be done intensively, and we're prepared, to some extent, to sacrifice it as virtually a factory for the production of food, and not try to expect it to do everything. But in the meanwhile, make sure that the rest of the landscape is somehow or other compensating for that area which is intensively working for us.[6]

Kay Hull, federal member for Riverina, suggests that the human activity of irrigated farming should be imaginatively and physically divorced from river systems. 'Let us take our production water out of the debate about the health of the Murray-Darling Basin', writes Hull. The politician offers a striking denial of human embeddedness within natural terrains and our dependency on natural processes, suggesting that irrigation can operate independently of complex river ecologies: 'We must pipe from the dam wall away from the Murrumbidgee River completely!'[7]

As global population rises, some people believe genetically modified plants will enable more production from existing farmland, saving 'natural' terrains from the spread of agri-*culture*. 'Without biotechnology the world will need to clear more forests and wildlife habitats to keep food production balanced with rising populations', Nobel Prize winning agricultural scientist Norman Borlaug argued during a recent visit to Australia.[8]

On the southwest slopes, bands of revegetation plantings and patches of remnant bushland severed by fence lines from much wider farmland spaces remain the norm. 'Natural' fragments and farmland paddocks are rarely seen and managed as one. Export-oriented, industrial farming relies on processes of division. 'The fragmentation of components of the farm ecosystem and their integration with distant markets and industries is a characteristic of modern "scientific" agriculture', writes ecologist Vandana Shiva.[9]

Industrial production denies the ultimate dependence of farmland on ecological connectivity and biological diversity. Across rural Australia, notes landscape ecologist John Williams, it is often assumed that biodiversity is found only on conservation reserves, on uncleared agricultural land, or on remnant patches of bush on farming land that may or may not be fenced off. However, biodiversity in the agricultural and pastoral ecosystems that make up these lands is often central to the lands' productivity. Agriculture is an ecological enterprise that depends on ecosystem processes and functions – such as soil formation, nutrient cycling, maintenance of hydrological cycles, pollination of crops – which are driven by interactions between elements of biodiversity.[10]

Farmland resilience, stability and enduring productivity depend on the presence of diverse and interactive communities of species. Inside paddocks devoted to imported crops and pastures, elderly and isolated paddock trees are habitat for birds, insects, reptiles and small mammals. These creatures consume and contain pest insects. Goannas and carpet pythons eat mice and rabbits, but require old trees and logs. Deep roots of trees, shrubs and perennial grasses draw leached nutrients back to the surface and lower salty watertables. Paddock trees allow woodland bird and other animal species to move between wider remnants and survive. Areas of regenerating native vegetation improve soil structures, limit erosion and minimise acidification. 'Sustainability and diversity are ecologically linked', explains Shiva,

> because diversity offers the multiplicity of interactions which can heal ecological disturbance to any part of the system. Non-sustainability and uniformity means that disturbance to one part is translated into a disturbance to all other parts. Instead of being contained, ecological destabilisation tends to be amplified.[11]

Dust storms rose from the slopes and plains of southern New South Wales during the droughty spring of 2002, from paddocks eaten bare. Brown-red clouds rolled east across the tablelands. More farmers should adopt alternative agricultural methods better integrated with local ecologies, some people argued in response to the dramatic erosion event. Even during droughts, they suggested, primary production could operate differently. Others pointed to descriptions by explorers of dust storms sweeping the region well before the arrival of graziers and wheat farmers.

In his book *Dust bowl*, Donald Worster presents a moving narrative of ecological and social catastrophe on the inland plains of America in the 1930s. Widespread denial of natural connectivity generated the crisis, the environmental historian argues, an observation equally relevant to the past and present of Australian farming. 'Nothing was fixed or permanent; man did not come into a perfectly stable or finished world on the plains', Worster writes. However,

> he did encounter there a set of alliances that might have helped him survive. All the living things needed each other, depended on each other, to withstand the harsher side of climate. The earliest humans to settle in the region understood that interdependency well, and respected it, but the white man did neither.[12]

Dominant systems of logic based on simplistic dichotomies underlie a general failure to see farmland as part of wider natural systems. Strategies to gain mastery over rural places and deliver primary produce to export markets require perceptions of farmland as a uniform 'Other', an inferior domain distinctly separate from humanity. Colonial processes bring denial of supportive functions offered by diverse biological communities.

Philosopher Val Plumwood's understandings of power dynamics active across colonised terrains offer insights into the culture of industrial farming. For the powerful 'One', Plumwood explains,

> dependency on the Other cannot be acknowledged, since to acknowledge dependence on an Other who is seen as unworthy would threaten the One's sense of superiority and apartness.[13]

Notions of a fundamental divide between land and people block possibilities for the integration of agriculture into living systems. Folding primary production into complex and dynamic terrains requires acceptance of human embeddedness in nature, and comprehension of relations binding paddock spaces to wider ecologies. Dependence on insecticides, herbicides and inorganic fertilisers reflects a denial of human embeddedness within nature. Imagining a sharp divide between nature and culture brings disorder to local ecologies. The rise of responsive, mutually nourishing relationships between people and rural places is disabled. Understandings of agri-*culture* inevitably 'impacting' on an external *nature* justify the dominant order and promote efforts to limit our 'ecological footprint'.

Perhaps closer to reality, we can instead see industrial interventions in natural systems as destructive and misguided. Alternative perceptions of humans as responsive members of biological communities enable people to become agents of ecological connectivity and wellbeing, not fragmentation and destruction.

Beyond garden fences

One spring weekend we decided to meet for lunch in a homestead garden opened to the public, west of Wallendbeen. The driveway wove through paddocks of wheat and pasture to a house and garden on the crest of a hill. Inside the garden fence, visitors disappeared down gravel paths to consider European trees and rare Asian shrubs. The property owners had spent much time and effort building and maintaining the garden around the homestead. Apricot roses and pastel irises jostled for attention inside curved garden beds. Blossoming spikes of purple echium swayed in the warm breeze. A pecan tree broke the uniformity of a wide lawn. Shrubs heavy with flowers framed views of farmland and black Angus cattle.

I remembered camping with a school group in a state forest nearby, where ironbark trees and different varieties of acacias thrive. Driving to visit the homestead garden, I noticed spear grass and native cherry growing on the roadside reserve. Yellow box stood near the creek flowing northwest towards the Lachlan River. Apart from a handful of white box trees, centuries old, I saw no plants inside the garden fence indigenous to the district. Each specimen, carefully tended, came from elsewhere. It appeared that the creators of the homestead garden had resisted sensual integration with the lively patterns of hillsides and creek valleys beyond the garden fence. The manicured site seemed an outpost of somewhere else, of the distant places and markets the western slopes and plains have served since inland colonisation began.

Gardens encircling family homes hold diverse meanings and associations. An array of personal styles and memories underlie the choices and practices of domestic gardeners. When settler gardens and gardening traditions are considered within broader contexts of western culture and colonial history, other meanings and purposes are revealed. The planting and fencing of gardens held particular significance to early Australian colonists and to British settlers elsewhere. In contrast to the ceremonial and legal practices of other European powers seeking ownership of distant lands, historian Patricia Seed argues, a garden enclosed by a fence itself signified possession to the British.[14]

'Like most of our population, our gardening traditions and styles are largely imported', notes Peter Watts, Chairman of the Australian Garden History Society, in *The Oxford companion to Australian gardens*.[15] Since the start of European settlement in Australia, gardens based on British gardening styles have dominated public and domestic spaces. Some early colonists imported and nurtured European garden plants because the specimens evoked memories of loved places left behind. But as the proportion of settlers born in Australia climbed, introduced shrubs and deciduous trees no longer served the homesick. Ordered gardens symbolised the success of British colonisation. Public parks and domestic gardens planted mainly to exotic species proved that Australia was more than a vast sheepwalk. Signifiers of agriculture and commerce, manicured gardens suggested that settler Australians rightfully held the continent. Later in the nineteenth century, as squatters and selectors competed for land, ordered domestic gardens of mostly imported plants reflected rival claims over the same terrains.

In colonial Australia, women built most domestic gardens. During the early decades of rural colonisation, historian Paul Fox observes, homestead gardens were 'a frontier space where women created a space beyond the sheep run and animal husbandry of their menfolk.'[16] On one side of the fence, around homesteads, women tended beautiful and comfortable spaces. On the other side, men applied industrial technologies to battle fire, drought, pests and weeds, and to grow fat lambs and wheat for distant markets.

Throughout rural Australia, gardening styles and traditions imported from Europe continue to shape homestead gardens. Australia's Open Garden Scheme first opened rural and urban domestic gardens to the public in 1987. The non-profit organisation promotes the 'knowledge and pleasure' of gardening 'by opening Australia's most inspiring private gardens to the public.'[17] On the southwest slopes, a number of gardens surrounding the homes of relatively prosperous farming families are opened each year as part of the Open Garden Scheme. One garden considered inspirational encircles a homestead between Young and Cootamundra:

> Profuse plantings of climbing roses and perennials are complemented by unusual trees including a pair of thriving European limes. The glorious pink climbing rose 'Mme Griegoire Staechelin' rambles along the front verandah, and many David Austins also flourish.

Nearby, another homestead garden selected for public viewing features 'sweeping lawns' and an array of exotic plants:

> In the flower garden lychnis, catmint, hollyhocks, knautia, grasses and sedums feature along with bush roses. Six wooden obelisks support larger-growing roses such as 'Complicata' and 'Souvenir de la Malmaison'. A rosemary-hedged herb and vegetable garden and a quince walk are developing.

In the nineteenth century, Australian homestead gardens of European plants gave sanctuary from land seen as uncomfortable and inhospitable. Western agricultural methods and continual grazing intensified the effects of drought. As indigenous plants and animals vanished and local ecologies faltered, Paul Fox notes,

> the homestead garden came to be conceived of as an ideal world standing apart from the surrounding landscape; a place of refuge from the elemental forces of nature that periodically crippled the productive landscape.[18]

Homestead gardens became representations of elsewhere, of a distant, favoured terrain. Colonists of other lands and people pursue a project of assimilation, Val Plumwood explains, whereby they

> remake the colonised and their space in the image of the coloniser's own self-space, their own culture or land, which is represented as the paradigm of reason, beauty and order.[19]

Gardens made by settler Australians often reflect the assimilation project Plumwood describes. Domestic gardeners do battle with local ecologies to impose an imported, usually European aesthetic. Successful attempts are honoured. Lynne Landy, wife of Victorian governor John Landy, recently launched the fifteenth season of Australia's Open Garden Scheme at Government House in Melbourne, a venue steeped in colonial symbolism. She encouraged people to visit open gardens. 'For the gardens in the Garden Scheme are real gardens', Landy declared,

> developed by real people who have to struggle against rabbits, possums, soil problems and water shortages yet still manage to produce acres of roses, islands of colour and tranquil plots that soothe the soul, often designed with great artistry. I'd like to quote from the current issue of *Gardens Illustrated*, where Mike Calnan who is the Head of Gardens for the English National Trust says: 'We have a problem with our attitude towards gardeners. They are creating works of art and we should respect and reward that.' And the Australian Open Garden Scheme does just that.[20]

A pamphlet issued by the Riverina Regional Development Board invites tourists to discover 'magnificent heritage properties with their manicured gardens of roses and camellias bordered by golden fields of canola and wheat'. Fences between rural domestic gardens and production paddocks are physical and imaginative divisions crucial to the operation of industrial agriculture. More so than in other parts of western Europe, in England garden fences played a significant role in perceptions of nature and society. Separating 'wild' domains from 'cultivated' spaces, fences helped construct the simple binary of 'wildness' versus 'civilisation'.[21]

In Australia, gardens established by English colonists reflected the same beliefs. Settlers called land tended by Aborigines 'Wilderness', poet Judith Wright observed, 'hostile country'. The newcomers cast places unmarked by colonial activities as 'untamed, unpleasant and unproductive land – the Waste'.[22] Outside garden enclosures resided savage forces, agencies requiring subjugation. The production of food and fibre in paddocks beyond garden fences, it was assumed, required strenuous effort to quell and erase the troublesome forces of the land.

Inside garden fences, around homesteads, are comfortable spaces. Soft lawns, garden seats, shady trees and sheltered spaces meet human desires. Beyond domestic gardens, across open paddocks, land and natural forces are subjected to industrial and commercial demands. Despite the revegetation work of Landcare groups and farmers, most paddock spaces remain windswept, exposed to sunlight, reshaped by machinery.

Garden fences reflect and maintain the dominant perception of a sharp divide between people and land. Fences enable an imaginative process Plumwood calls 'hyper-separation'.[23] Continuities between land and people are denied, differences exaggerated. Rural places and other species are seen as devoid of expressive individuality and agency, a subordinated and homogenous 'Other' vastly different from humankind. Modern agriculture devalues and neglects natural connectivity and dynamism. Colonists divorce themselves from farmland, grasping control of terrain seen primarily as a 'natural resource' available for industrial use and commercial gain.

By casting a sharp division between people and farmland, the imaginative framework of modern agriculture enables the diverse nature of rural places to be denied. Simple homogeneity, not complex diversity, is seen to characterise farmland. Standard demands from vast populations require the standardisation of productive rural places. To meet market demands for standardised food and fibre products, settlers must impose a uniform aesthetic of wide crops and pastures over varying rural terrains. Particularities and intricacies of places are erased.

'Bread is the staple food of the white races' observed Robert Watt, Emeritus Professor of Agriculture at the University of Sydney, in 1955.[24] In recent centuries across the globe, European colonists generated an aesthetic sameness as they applied industrial technologies to harness land for wheat production. Surnames and languages from western Europe today dominate the temperate grasslands of all the continents, in each the terrain favoured by wheat.[25]

One afternoon in the National Library I leafed through *An open land: photographs of the midwest, 1852–1982*.[26] Inside the book, a photo titled 'Grain elevator and plowed fields, Wellington, Kansas, 1973' shows a paddock of stubble and turned earth. Grain silos stand beside railway tracks. Several mature trees are visible on the horizon.

I searched 'wellington kansas' on the web to discover more. The first website showed a photo of a road sign. 'Wheat Capital of the World', read the sign, below a painting of three grain-filled heads of wheat. The two photos from the agricultural heartland of the United States are similar to some I have taken on the south-west slopes of New South Wales. Mine too show wide paddocks of ploughed earth, concrete silos, railway tracks, elderly trees and road signs representing districts in terms of wheat production.

Perceptions of rural places and other species as homogenous and inferior enable the harnessing of land for primary production. Powerful, monological relations require devaluation and denial of individuality and diversity in a subordinated group. The dominant

'One' sees the colonised and distinctly different 'Other' in simple terms, as readily knowable and controllable. Responsiveness towards particular needs of local ecologies only complicates and blocks attempts to secure wide areas for monocultural, export-oriented production, for the generation of standard products demanded by distant markets. Power enables colonists to ignore expressions and needs rising from subordinated places, to avoid entering dialogue, as Val Plumwood explains:

> Notice how these features result from power and work together: thus, to the One, sensitivity to differences among the Others is of little importance, unless they affect his own welfare, because power or force can take the place of sensitivity, whereas sensitivity to differences among the masters is likely to be very important for the survival of the subordinated. Diversity which is surplus to the centre's desire and need does not require respect or recognition. Thus knowability and lack of diversity is likely to be strongly stressed for the subordinated group.[27]

Drought and fire shaped the living systems of the southwest slopes. Regardless of the season, many grassy woodland plants respond to rain with flowers and fresh growth. In preparation for dry times, perennial grasses and shrubs send roots deep down. The fleshy tubers of chocolate lilies, native geraniums, waxlip orchids and other forbs enable regeneration after drought and fire.

Instead of recognising natural diversity and integrating primary production within differing local ecologies, settlers applied western science and industrial technologies to impose a simple patchwork of imported crops and pastures. Farmers rely on the successful harvest of shallow-rooted annual crop species – wide monocultures vulnerable to insects, diseases, frosts, storms, droughts, bushfires and other natural forces. In contrast to the opportunistic habits of native plants, imported crops depend on regular seasonal patterns. Colonial denial of both ecological diversity and the erratic patterns of Australian climate continue to block cultural changes necessary for the regeneration of rural places. Refusal to recognise and respond to particular natural patterns enables the operation of industrial systems of grazing and cropping.

'Cootamundra is a place where the seasons are distinct', claimed a garden journalist after touring local gardens in 1992.[28] The magazine article cast exotic garden plants and regular European seasonal patterns as natural realities of the southwest slopes. 'Spring seems to be dominated by forsythia and crab apples', one homestead gardener told the journalist,

> summer is just a mass of old-fashioned roses that I have been growing for 25 years. Autumn sees all the berries come out and the garden is full of clumps of maples that turn a wonderful colour. Winter has camellias in bloom and lots of bulbs popping up.

Wiradjuri artist and writer HJ Wedge was born on Erambie mission, an Aboriginal reserve beside the Lachlan River at Cowra, northeast of Young.[29] In his book *Wiradjuri spirit man*, Wedge describes a garden different from those established by colonists around homesteads in Wiradjuri country.

Wedge doesn't perceive sharp boundaries between domestic spaces and productive land. Before colonisation, he explains, the entire continent was a garden, a 'nourishing terrain' in which people saw themselves as inextricably embedded.[30] By the late twentieth century, the bountiful garden was lost:

> Over the centuries many children have been living in this garden of theirs with their own lore. Till the white people came along and destroyed the garden over two hundred years. They are destroying the rainforests to make toilet paper and digging up the land to mine and poison the land and water with chemical waste and if you don't fuckin' believe me you can get in your car and drive around this ruined garden of ours.[31]

Ethical behaviour, suggested Aldo Leopold, rests on a single premise: 'the individual is a member of a community of interdependent parts.'[32] Uncooperative behaviour is unethical, because it undermines the strength and wellbeing of interconnected social and ecological communities. Western systems of knowledge and colonial processes continue to block possibilities for both urban and rural Australians to accept and understand their own physical embeddedness within the natural patterns of farming regions. A denial of human dependency on living systems underlies neglectful actions. Perceptions of people as separated from and in control over farmland, and failure to acknowledge and respond to the needs of rural places, perpetuate the wounding of local ecologies throughout the agricultural heartlands of Australia.

An industrial picturesque

A giant kurrajong stands in Albert Park, beside the railway line at Cootamundra. Sturdy limbs hold cool green sprays of leaves. Enormous roots swell the red earth into a wide mound. Nearby rises a yellow box, elderly and tall, a trunk of flaky bark. Most trees and shrubs in the park come from elsewhere. English oaks shade play equipment. There are elms and pines, peppercorn trees and oleanders.

The kurrajong and yellow box probably grew to maturity a long time before 1878, the year Cootamundra townspeople chose a name for the local recreation ground. They named Albert Park in memory of Prince Albert, husband of Queen Victoria, a brass plaque beside the cricket oval explains.

The arrival of the Great Southern Railway the year before had enlivened the town. Residents decided to transform the unkempt public recreation ground into an ordered park. The space would enable 'recreation to help in the physical development and rational enjoyment of our youth', observed the *Cootamundra Herald*.[33]

Many people were expected to settle in Cootamundra. They would need somewhere to walk, breathe fresh air and rest. The park's trustees fenced the five hectares between the railway track and the main street. In winter they planted trees ordered from

Sydney. Parkland was harrowed and sown to couch grass. As spring approached, Albert Park would 'assume a really beautiful and ornate aspect', the *Herald* told readers.

Cootamundra grew as closer settlement and farmland spread across the slopes in the last decades of the nineteenth century. In 1900 a Sydney journalist described a 'fine town, with a nineteenth century air of civilisation about it'.[34] The setting of the town appealed to the reporter:

> The situation is a most charming one, the town being picturesquely situated, though not too closely, near wooded hills to the south, one of which, Mount Coghlan, the highest, rises into a minor kind of sublimity, the sides clothed with forest trees.

The journalist perceived 'a minor kind of sublimity' in Mount Coghlan, a steep granite hill on the southern side of Cootamundra. In the nineteenth century, descriptions of 'sublime' natural features, rugged and monumental, implied the presence of awesome natural forces able to challenge and overwhelm human will. Transcendence and excess categorised certain landscapes as 'sublime'.

Describing the town as 'picturesquely situated', the reporter framed the Cootamundra district in 'picturesque' terms, another popular aesthetic category of the nineteenth century. Places described as 'picturesque' were considered distinctly mild and useful to humanity. Applying the perceptual framework of the picturesque, the visiting reporter cast the undulating, grassy woodlands surrounding Cootamundra as fertile terrain available for transformation into farmland. 'Where weeds formerly throve in rank profusion', wrote the journalist, 'apple and pear and peach trees are now heavy with precious fruit'. Before the arrival of the railway, the area around the town of Cootamundra 'yielded a subsistence to a few'. Now cleared of trees and dramatically altered by industrial technologies, the same land 'repays the skilful farmer a hundredfold'. Rail access to Sydney and Melbourne secured a market for local grain.

As selectors transformed grassy and timbered hillsides into farming paddocks, much land unfortunately remained beyond reach inside the boundaries of wide pastoral stations:

> All over the district farms have been laid out and paddock after paddock of wheat, thousands of acres in extent, have been cultivated. Still a great portion of the land is locked up in principalities. This country produces cereals in a state of perfection. The wheat raised here is of the best quality for flour.

In the eighteenth century, William Gilpin identified as 'picturesque' any natural elements and objects considered 'proper subjects for painting.'[35] An influential authority on the new and popular aesthetic style, the Englishman felt that landscape paintings must conform to certain conventions to be categorised as picturesque compositions.

A picturesque landscape painting required a foreground, middle ground and background. An interesting and unifying visual feature occupied the middle ground. Towards the edges of the picturesque composition, trees or rocks darker than the middle ground framed the work and drew attention to the central feature. Picturesque conventions never demanded close and faithful attention to unique patterns of the land. Artists of the picturesque considered vistas and places as resources available to the human aesthetic imagination. Picturesque artists traditionally captured images of land from an elevated position, literary critic Jonathan Bate explains, from 'a raised promontory in which the spectator stands above the earth, looking down over it in an attitude of Enlightenment mastery.'[36]

The aesthetic category of the picturesque emerged in Britain as the industrial revolution unfolded in the eighteenth and nineteenth centuries. Picturesque conventions placed humans above and in control over places and nature, reflecting and reinforcing the Enlightenment philosophies that enabled industrialisation. The framework of the picturesque presented wide areas reshaped or reimagined by humans as natural and unchanged, obscuring histories and particularities. Picturesque representations distanced people from places, enabling subjugation of land, other species and natural forces for economic gain.

Convict artist William Buelow Gould is the central character of *Gould's book of fish*, a novel by Tasmanian writer Richard Flanagan. Gould disparages picturesque landscape painting, a genre admired by his masters and other powerful colonial officials:

> I care not to paint pretend pictures of long views which blur the particular & insult the living, those *landscapes* that trash the truth as they reach ever upwards into the sky, as though we only know somewhere or somebody from a distance – that's the lie of the land while the truth is never far away but up close in the dirt, in the vile details of slime & scale & filth along with the Devil, along with the angels, & all snared within the earth & us, all embodied in a single pulse of a heart – mine, yours, ours.[37]

Reflecting on the writings of European explorers in Australia, cultural historian Simon Ryan reveals particular ideologies within picturesque conventions of seeing. Picturesque aesthetics, Ryan argues, hide

> an instrumentalist agenda which establishes nature solely as an object to be valued according to its ability to please and serve human beings.[38]

Australian explorers traversed the continent primarily to find rivers and fertile soils – terrain suited to pastoralism and agriculture. Imperial desire for natural assets underlay descriptions by explorers of certain views and places as picturesque. As Ryan explains, 'if the land was picturesque it was ripe for transformation into wealth.'

On the southwest slopes of New South Wales, explorers frequently described fertile, grassy terrain in picturesque terms. Charles Sturt and his exploratory party journeyed down the Murrumbidgee valley west of Gundagai in December 1829. Sturt climbed a hill and described the view:

> There is an extensive flat to the westward, which we shall traverse tomorrow, in which direction the valley of the Morambidgee is plainly marked out. The river has increased both in depth and breadth, and rolls along a vast body of water. The general appearance of the country is unchanged. It is extremely rich and picturesque.[39]

Sturt neared the future site of Wagga a few days later. He thought the area held great potential for British colonisation, and again applied picturesque conventions of seeing to frame his judgement:

> The forests behind the flats in this branch of the river are admirably adapted for grazing. The soil has fallen off a little in quality, but it is still fine, and is better perhaps for the purposes of agriculture than the richer earths of the upper district. We experience delightful & cloudy weather, and the scenery around us is cheering and picturesque.[40]

In Australia, Ryan observes, British settlers adapted the strict picturesque conventions identified by Gilpin to satisfy agendas of colonisation. Modified picturesque frameworks gave a pleasing sense of approval to colonial activities and alterations. According to Gilpin, picturesque landscapes never featured evidence of industrial activity. Australian explorers altered the picturesque framework, sanctioning British colonisation by placing buildings and fences within picturesque descriptions and compositions. The 'industrious hand of man', explorer John Oxley observed, was 'improving the works of nature'.[41]

Picturesque frameworks still dominate representations of rural Australian places. Wire fences and even gully erosion are cast as 'natural' components of the land. 'We're as much a part of the Country as the Gums and the Gullies' asserts an advertisement for steel agricultural products.[42] Naturalisation of paddock fencing and other industrial devices suggests that exclusive possession of land and agendas of mastery over rural places are natural also.

Cootamundra Shire boundary sign,
Harden road.

A sign erected by the Riverina Regional Development Board stands beside the road on the Cootamundra Shire boundary. The sign welcomes travellers to the shire, 'Gateway to Riverina ... naturally'. According to its website, the board chose the slogan 'Riverina ... naturally' to emphasise

> the clean air, water and soil of the Riverina, which support the rich and diverse agricultural and horticultural production underpinning the regional economy.

Positioned high on the metal sign is the board's logo, a wide and stylised view of a crop. Straight green lines representing crop rows narrow then disappear over the horizon. The rising sun radiates into blue sky, beams reflecting the crop lines below – a harmonious blend of sun, earth and industry.

Presenting industrial production as beautiful, the logo reflects what political scientist James Scott calls 'the visual aesthetic of agricultural high modernism', a style generated by a confident and expansive industrial culture.[43] There is no tension, the logo suggests, between local ecologies and the intense demands of an export-oriented, industrial system of primary production. The view is uniform and picturesque, a pleasing and comfortable prospect.

Inscribed below the logo, the board's slogan, 'Riverina ... naturally', suggests the dominant western perception of land as a 'natural resource' available for transformation into regular patterns by industrial machines. Despite the claim made by the slogan, wide paddocks groomed and planted to single species cannot operate as integrated and interactive parts of complex natural systems. Irregularities and complexities of rural places are obscured and denied by the logo and slogan, reaffirming a way of seeing responsible for the disordering of local ecologies.

Other applications of the picturesque fuel commercial activity in the Cootamundra district. A tourist brochure issued by the Cootamundra Development Corporation suggests various scenic drives. Visitors are directed to an 'excellent photo stop' on a rise northwest of Cootamundra, where 'the views out towards Stockinbingal are nothing short of spectacular'. From the vantage point tourists may 'soak in the scenic farmland views'. As the main road southeast of Cootamundra rises, 'magnificent views across the Muttama Valley' become visible. The outlook from the ridge is 'spectacular all year round, particularly in September – a patchwork of green and yellow with the Canola in flower.' According to the promotional pamphlet, local farming practices are comfortably intertwined with natural cycles. Shifts in industrial activity mark seasonal changes, an image serving to naturalise the forceful imprint of agricultural science and technology on the land:

> Tractors preparing paddocks for sowing in the autumn; new crops in a range of vibrant colours and fruit and nut trees in blossom herald spring's bounty; and towards the end of the year comes the harvest when everyone works around the clock. That's when you'll see headers stripping crops till late at night, their headlights

moving up and down the rows like spaceships in the dark. It's also when you'll find truckloads of grain being carted to the silos.

To maintain the high levels of primary production necessary for farmers to survive in the global marketplace, land must be continually reworked. Green paddocks fade swiftly in preparation for cropping as broad-spectrum herbicides kill every grass and forb. Wide ploughs and seeding equipment expose and crumble soils. Diesel merchants prosper during sowing and harvest as tractors work paddocks and trucks haul fertiliser and grain.

Industrial agriculture uses picturesque frameworks to cast land as malleable terrain onto which uniform crop monocultures may be imposed. Typical rural scenes illustrate a calendar published recently by the Prime Wheat Association of New South Wales. One photo shows a view of Harden district farmland. Rows of cereal crop stubble extend across a slope behind a wire fence. True to the formal conventions of the picturesque, the shaded branches and trunk of a yellow box tree frame a bright expanse of stubble in the middle ground. A second photo shows a tractor driving through crop stubble swept by flame, across land undergoing strident transformation. Another shows two children gazing into a yellow paddock of flowering canola. Beside the golden canola crop, at the centre of the photo, a sickly tree appears close to death.

The image of the two children, seemingly oblivious to the dying paddock tree, suggests the ideological agenda of the picturesque. To enable mastery over farmland, picturesque imagery constructs perceptual and emotional barriers between people and the intricate natural patterns of rural places. Concern for a sickly paddock tree is unwelcome, as attention may undermine the validity of the industrial cropping practices responsible for such ecological disorders.

Picturesque frameworks of perception obscure and deny the fragmentation of natural connectivity wrought by modern farming. Industrial farmland is presented as attractive and wholesome, undermining potential criticism. Sites presented as 'landscape', philosopher Stephen Muecke observes, 'are often conceived as points of consent between nature and culture, where homeliness billows forth in a dream of harmony'.[44] Contestation is hidden.

Distance between people and land engendered by the picturesque enables the unleashing of force. 'Have more power to the ground than ever before', declares an advertisement published recently for a new range of tractors.[45] 'The powerful 8.1 litre and 12.5 litre POWERTECH® engines create a tremendous power response.' Inside the tractor cabin, high above the paddock, operators are in control:

> The exclusive CommandView™ cab helps make hours fly by. It's quiet, spacious and visibility is second to none. Controls rest virtually at your fingertips with the exclusive John Deere Command-ARM™. The all-new FieldVision™ lighting option uses high-intensity xenon-gas lamps to improve vision, clarity and distance.

~ ~ ~

From a stony hillside, under the dense shade of an old kurrajong tree, I gaze towards blue ranges defining the horizon. Australians are familiar with images of rural vistas like the one extending before me. Where the slope levels, homestead chimneys and a corrugated iron roof rise from a wide garden. Ewes and lambs graze ryegrass and clover beside tall eucalypts. Several willow trees stand on the bank of a creek, near shearers' quarters and a voluminous woolshed. A steel fence passes a weatherboard cottage and a muddy dam. Sheep tracks weave outwards from a cement water trough, across a paddock threadbare with loose stubble.

The remains of a racehorse once nourished the elderly kurrajong now shading me. 'Featherstitch 1905–23', a marble headstone reads beside the tree. My father's grandfather bred racehorses here on Retreat, a property between Cootamundra and Temora.

Horses grazed these hillsides long before my family came. Pinchgut Creek curves through paddocks below, past old eucalypts towards the Murrumbidgee River. In the early decades of colonisation, horses escaping riverside squatting runs retreated to distant hills. Great mobs of wild horses alongside Pinchgut Creek gave Retreat its name. An elderly resident of the Temora district remembered Mimosa station, west of Retreat, 'infested with wild horses and kangaroos' in the 1870s.[46]

About the same time, men trapped and killed in one year twelve hundred horses on Berthong station, northeast of Retreat.[47] My grandfather told my father why a paddock on Retreat was named 'Trapyards'. Years ago there, station workers corralled mobs of wild horses inside a set of yards. The animals trapped, a man stood on either side of a narrow gate, each holding a pole with a sharpened blade from a pair of hand shears bound to the end. When the gate opened, the men slashed the bellies of the horses as they rushed to exit. The frightened animals galloped away to die slowly among white cypress pines and yellow box trees. Graziers across the region considered mobs of wild horses a demonic barrier to pastoral development, Mary Gilmore explained, 'to be blotted out like the blacks.'[48]

Several quandong trees, my great aunt remembers, grew in a paddock south of the hillside where I stand. And a pair of bush stone-curlews nested beside a road near the woolshed. The quandongs and curlews vanished decades ago. Southeast from my vantage point, past a dark green paddock of lucerne, I see a tall yellow box, pale limbs dense with leaves, rising above the woolshed and its network of mesh yards. My father taught me to ride a bike there, on red earth compressed by generations of sheep, hooting encouragement as I began to circle the wide, fibrous girth of the elderly eucalypt. Beneath the kurrajong tree, I view the childhood event years later, from a distant position. As in familiar representations of golden rural vistas, particular details remain obscure.

I leave the cool shade of the kurrajong and walk downhill. Over the road in the lucerne paddock, pale green domes of *division* (145)

peppercorn trees ripple in the heat. The original Retreat homestead stood somewhere between them, above a swamp and several dams on Pinchgut Creek. Traced in black ink, the swamp appears on Parish of Hurley maps issued by the Department of Lands more than a century ago.

Where old maps indicate swampland I find a deep, dry creek channel curving through crop stubble. No swampland plants, no expanse of dark, wet earth. The paddock upstream turns sodden when winters are especially wet. There, east of Retreat homestead, Pinchgut Creek loses definition in a level area studded with rushes. My great aunt remembers walking with her brothers and sisters to school, a small building on the other side of the flat and sometimes swampy paddock. To save her feet soaking when the area was wet, she carefully walked right alongside the fence where the ground was slightly higher. Her parents came to Retreat a century ago and planted an orchard below their new pisé homestead, on the northern edge of the swamp marked on parish maps, into earth deep and moist.

Only the upstream part of the swampland remains today, where elderly fruit trees grow. Downstream, in the paddock of crop stubble where the main body of the swamp once lay, I step into the dusty channel of Pinchgut Creek. Several layers, I notice, make up the channel wall. From the paddock surface down is a deep band of light coloured soil washed from surrounding hills. Underneath, lower down the channel wall, runs a dark layer of ancient swamp earth, rich with rotted plant material. The soil bands record a history of cultivation, grazing and drought, of hillsides disturbed and bare.

In summer heat, questions rise from Pinchgut Creek. When did heavy storms bury the swampland under loose soil? How long did the channel take to carve through the earthen layers? Years ago someone, perhaps my grandfather, threw old fencing wire, bent steel posts, broken machinery parts and rusted iron sheets into the eroding head of the channel upstream to stem the loss of soil. But floodwaters proved stronger. Erosion continues today, as

Pinchgut Creek channels into the northern remnant of winter swampland, slowly erasing the natural feature.

In the creek bed lies a sheep carcass – rotten skin, bones and wool – in a dry waterhole below the black band of swamp earth. Midday sunlight makes its white skull shine. I climb out, onto the overlay of eroded soil and stubble, the remains of a wheat crop. A willy willy approaches from the southwest, where Pinchgut Creek vanishes behind a slope. The spinning tunnel of dusty air and stubble whispers vigorously through the crowns of yellow box trees. On the other side of the waterway, near where the parish maps say the swamp began, a tall and sturdy red gum stands alone. The tree looks like a river red gum, a species common in watery places, but rare in dry paddocks like this one, beside a minor creek high in the Murrumbidgee basin. I imagine the tree growing to maturity centuries ago, when rushes grew here in damp black earth. Does the ancient organism, I wonder, hold memories of the dead swamp and its people?

Walking away, I stop to examine flakes of stone near the creek bank, tools and workings unearthed by plough blades and the hard hooves of livestock. Some flakes are long and narrow. Others, knocked from white quartz, are short and angular. I admire a large flake, smoky blue with red lines. My finger traces the sharp edge of a blade worked from a smooth, fine-grained stone, pure black. Squatting, close to the earth, I encounter traces of history – stories and particularities invisible from a distance. A call on the present seems to rise from the lost swamp. Paddocks silenced, emptied of history and varied life, are made to grow uniform products for distant markets. In the summer dryness, the wounded place cries out for recognition and response.

Curtains of sandstone and power

'The Riverina, as the country watered by the Murrumbidgee is called, presents the most classical landscape in Australia', historian and novelist Marjorie Eldershaw wrote in 1939. At the time,

Australian nationalism was bound to farmland and agricultural production:

> In the paddocks there are all the tones of sunburnt yellow, brown, sage green, an undercurrent of purple and mauve, in the soil, in dead bushes and timber. Distant scrub and shadows are blue. Fallowed land and lucerne crops make patches of rich dark colour and vivid green. The wheat stands green and bronze. Sheets of water in the dams shine like galvanised iron, iron roofs look like water among the trees. Breakwinds of dark trees round bleached white homesteads accent the country. The silos, great blind towers, standing at the railway sidings bring the scene into focus. The paddocks are scrabbled with sheep tracks as if giant fingers had been drawn through the grass. The sheep themselves, neutral coloured, fit the country like a natural feature. Because wheat and pasture mingle, the Riverina is fairly closely settled. If there isn't a homestead in sight, there's likely to be a share farmer's house with its pepperina or a hut beside a dam, or at least a fence and a gate, and at night those little squares of yellow light, so significant in the wide country darkness, are scattered far and wide like lonely stars.[49]

Late in the nineteenth century, as colonial leaders negotiated Federation, and throughout much of the twentieth century, Australians drew upon rural places and imagery to construct shared notions of character and identity. Artists such as Tom Roberts and Frederick McCubbin painted popular images of sunlit paddocks and bustling woolsheds, images to define and construct an emerging nation. Infatuated with sunlight, they reflected new, optimistic understandings of Australian places. The artists presented nature as relatively kind and amenable – 'even the melancholy which flavours the work of McCubbin has an affectionate glow to it', observes cultural historian John Rickard.[50]

When Sydney hosted the official launch of the Commonwealth of Australia in January 1901, organisers used rural imagery and themes to define the nation.[51] A grand procession wound through the city centre. Crowds cheered on Bridge Street as marching soldiers and carriages bearing dignitaries passed under a triumphal arch dedicated to the Australian wool industry. 'Welcome to the Land of the Golden Fleece', the archway read beneath a dome covered in scoured wool. The procession encountered a second arch celebrating the farmers of New South Wales and crowned by a model steel plough. 'Ceres welcomes the Commonwealth', an inscription on the archway announced, where a plaster Ceres, Greek goddess of agriculture, emerged from flags and wheat sheaves.

An enduring and distinctly Australian character formed in the nineteenth century from the experiences and culture of rural working men, historian Russel Ward argued half a century after Federation. According to Ward, inland colonisation generated the 'typical Australian', a knockabout bushman known for his mateship and democratic outlook.[52]

Differences in the cultures and realities of rural and urban domains have existed ever since the emergence of western agricultural systems and urban development millennia ago. In colonial Australia, and throughout the twentieth century, popular representations of rural places and inhabitants reinforced and magnified notions of difference.

Cultural historian Don Aitkin identifies the rise of an ideology he calls 'countrymindedness' in rural Australia before World War II.[53] Countrymindedness casts urban lifestyles as degenerate and individualistic. In contrast, farming and grazing are presented as noble pursuits. According to the ideology of countrymindedness, bureaucratic and government interference always threatens the dignity and freedom of rural landholders. As rural sociologists Lisa Bourke and Stewart Lockie observe, countrymindedness remains alive outside Australian cities, where the interests of big landholders are often presented as the interests of all rural dwellers.[54]

In the twentieth century, countrymindedness constructed a sharp divide between rural and urban interests, an imaginary separation growing as decades passed. The cultural, social and economic rift separating rural and urban Australians, historian Geoffrey Blainey recently observed, 'is wider than at any time in the last 150 years.'[55]

Changing settlement patterns broke ties between city and country in the second half of the twentieth century. Mechanisation, improved transport, declining terms of trade and greater exposure to globalised competition drove people from Australian agricultural regions. At the end of World War II, almost one third of the Australian population lived in rural areas. By 1970, only one sixth of Australians did so. Throughout the 1970s, 45 000 workers and 28 000 farmers left agriculture.[56]

Rural depopulation continued in subsequent decades. Government policies and economic pressures fostered an intensification of industrial farming practices and the expansion of average farm sizes. International migration packed even more people into cities, mainly along the eastern coastline.

Shifts in national consciousness and priorities undermined alternative possibilities. No longer was farmland a domain for people, a foundation for national identity and social stability. Instead, the rise of economic fundamentalist ideologies generated dominant perceptions of agricultural land as the material basis to an industrial system of production and export trade.

In the final decades of the twentieth century, Australian governments removed tariffs and other economic devices protecting farmers and farmland from the variability and intense competition of the global marketplace. Economic fundamentalism undermined and erased ethical frameworks of care and responsibility. Government policies shifted away from an emphasis on social equity, focusing instead on economic efficiency.

Transformations in government policy reflected and reinforced national and global patterns of change. To help farmland bear the demands of industrial, export-oriented agriculture, governments directed funding to Landcare projects and other remedial schemes. Few people argued that farmland represented anything other than a natural resource for modern agriculture, a successfully competitive global industry. Narrow perceptions of rural terrain as merely productive space banished history and people from farmland.

Sandstone ridges contain the sprawling city of Sydney across a coastal plain. As the twentieth century drew to a close, the term 'sandstone curtain' described perceptions of a deepening divide between city and country.

Rural and urban responses to mounting scientific evidence of ecological decline in agricultural regions have intensified the sense of separation. Sociologists Geoffrey Lawrence and Ian Gray describe a change in urban attitudes over recent decades.[57] No longer do many city dwellers honour the energetic efforts of farmers to harness land and produce food and fibre in a variable climate. Instead, the sociologists observe, people tend to blame farmers for land degradation.

Simplistic responses by city folk reinforce the sandstone curtain. Perhaps few urban dwellers understand the historical and cultural context to the ecological disordering of rural places, or accept the direct links between urban consumption, national export policies and ecologically insensitive systems of primary production.

Responses made by rural dwellers during debates about land degradation have also magnified existing perceptions of a country versus city dynamic. As environmental scientists and lobbyists push for changes to farming practices, landholder interest groups claim exclusive decision making and property rights over farmland. By constructing an oppositional, exclusive position for rural landholders, farming groups undermine what sense remains of common interests between rural and urban sectors. In a recent newspaper advertisement, the NSW Farmers Association issued a 'call to arms', rousing farmers to oppose 'misleading claims by extreme "green" groups in the city media'.[58] The organisation represents debates about rural ecological problems as a clash of separate interests:

> To counter these claims head on, the Association is collecting data to tell the real story – that farmers are responsible custodians of the land, practising environmentally sustainable farming methods.

NSW Farmers made no references in the advertisement to scientific evidence of ecological disorder, or to market dynamics forcing landholders to put immediate economic demands before the longer-term ecological needs of rural places. Instead of seeking opportunities to work with environmental groups towards structural changes and ecological regeneration, farming interest groups seek concessions and privileges by widening the rift between city and country.

In the nineteenth and early twentieth centuries, Australians honoured rural settlers as nation builders. 'Noble pioneers' took risks to secure and develop grazing and farming land not for personal gain, the legend went, but for the greater good. By the 1950s, the European conquest of Australia was complete. Rural people and places were no longer thought central to national identity. Australians looked beyond farmland to other terrains for definition of nation and self.

Popular culture reflected and helped construct new understandings of rural places and national identity.[59] In the past, successful films like *The squatter's daughter* (1910) and *On our selection* (1932) represented rural Australia as habitable and productive, a resource for national prosperity and identity. Later in the twentieth century, *Wake in fright* (1971), *Picnic at Hanging Rock* (1975), *The adventures of Priscilla: Queen of the Desert* (1994) and other popular films cast rural places as strange and malevolent.

In search of natural beauty and a new version of national identity, Australians turned away from farmland and towards regions relatively untouched by industrial society. In 1983, the Australian electorate voted the Labor Party into power, partly because it promised to save a forested valley of Tasmanian 'wilderness'. The new government prohibited the construction of a hydroelectric dam that would have drowned the rugged Franklin River valley.

Displays of national symbolism during the 2000 Olympic Games in Sydney demonstrate the popular association of desert and coastal environments with Australian identity. The passage of the Olympic torch across the continent began in the central Australian desert at Uluru, a place enclosed within a national park and often presented as the symbolic heart of the nation. At the Olympic opening ceremony in Sydney, more than five hundred performers presented 'Deep sea dreaming'. The elaborate aerial performance recognised a national 'love affair with the ocean', the organising committee explained.[60] Shifts in popular notions of national identity underlie the postwar ecological and social decline of rural Australia in complex ways. Late in the twentieth century, a devaluation of country life and terrains enabled the implementation of destructive economic and social policies. Agricultural regions became the subordinated, hyper-separated 'Other' to powerful economic interests.[61]

Today, rural people and places are cast as homogenous parts of an inferior and distinctly different domain, subordinate to global and urban marketplaces. The sandstone curtain obscures and enables modern systems of food and fibre production, distribution, processing and marketing. Urban consumers are pushed away from sites of primary production. When sharp divides exist between city and country, urban consumers cannot easily develop responsive, careful relationships with particular rural places and people. In *Fatal harvest: the tragedy of industrial agriculture*, farmer Wendell Berry describes an unfortunate dynamic of economics, ignorance and destruction:

> The global economy institutionalises a global ignorance, in which producers and consumers cannot know or care about one another and in which the histories of all products will be lost. In such a circumstance, the degradation of products and places, producers and consumers is inevitable.[62]

Les Murray charts a shift in the consciousness of Sydney dwellers away from rural people and land in his poem 'Sydney and the Bush'.[63] At the onset of British colonisation, Murray observes, there was only 'the Bush', an endless and uniform terrain beyond the new place of Sydney. Settlers erased natural patterns to secure land for a town:

> When Sydney and the Bush first met
> there was no open ground
> and men and girls, in chains and not,
> all made an urgent sound.
>
> Then convicts bled and warders bred,
> the Bush went back and back,
> the men of Fire and of Earth
> became White men and Black.

During the nationalist period of 'the bushman' in the late nineteenth and early twentieth centuries, Sydney residents held relatively close ties with rural people and places, Murray feels. Country and city interacted in mutually beneficial ways:

> When Sydney ordered lavish books
> and warmed her feet with coal
> the Bush came skylarking to town
> and gave poor folk a soul.

Later in the twentieth century, as people left Australian agricultural regions and coastal cities swelled with overseas migrants, a divide grew between rural and urban domains. City life became one of alienation and restriction:

> Then bushmen sank and factories rose
> and warders set the tone —
> the Bush, in quarter-acre blocks,
> helped families hold their own.

Murray describes extreme separation between urban and rural sectors as the twentieth century approached its end. Sydney residents saw themselves as entirely divorced from outlying rural areas:

> When Sydney and the Bush meet now
> there is antipathy
> and fashionable suburbs float
> at night, far out to sea.

Urban discourses helped construct the sandstone curtain. To maintain positions of power, Murray suggests, Sydney residents ridicule countrysiders, those 'Australians' representing the subordinated and anonymous 'Bush', the placeless rural 'Other' to the urban centre:

> When Sydney rules without the Bush
> she is a warders' shop
> with heavy dancing overhead
> the music will not stop
>
> and when the drummers want a laugh
> Australians are sent up.
> When Sydney and the Bush meet now
> there is no common ground.

Like the sandstone curtain, a product of oppositional discourse, the image presented by Murray of 'fashionable suburbs' floating 'far out to sea' speaks of sharp cultural divides between rural and urban people and places. Intimate physical ties binding the bodies of all urban consumers to rural places are obscured. Without knowledge of farmland and rural people, urban dwellers may comfortably evade ethical obligations to reciprocate the nourishment and wellbeing given by rural places. Sophisticated awareness of industrial farming systems and of the wider cultural and economic contexts of modern agriculture is blocked. Urban connections to rural domains are denied, the diverse voices and needs rising beyond cities ignored.

In earlier times, 'the Bush came skylarking to town and gave poor folk a soul', Murray writes. The imaginary sandstone curtain may impose significant costs indeed.

Sacred earth?

On the southwest slopes as summer approaches, eastern snake-necked turtles migrate from creeks and dams in search of watery nesting sites. Between Harden and Wallendbeen, I once saw a turtle stopped on double white lines in the middle of the road, curled inside its shell as traffic swept down to cross the bridge over Demondrille Creek. Pulling off the busy road, I waited for a semi-trailer to pass. The frightened creature didn't stir in my hands. I left the turtle on soft grass close to the waterway.

One spring afternoon years later, I drove south towards Wallendbeen from Young. Cherry orchards gave way to fat lambs and canola. I passed a property sign, 'Crookwood', inscribed in old English lettering. Sadly, the placename was fitting. On the other side of the road, sickly yellow box trees and dark tussocks of spike rush stood at the base of a slope in waterlogged, salty soil.

On the radio as I drove, musician Jimmy Little answered questions. He spoke quietly, with tenderness. His mother was a freshwater woman from Cumeragunga, Jimmy explained, a Murray River mission, and his father came from the coast. Jimmy took his identity from his mother's people. A freshwater man, he belonged to the Yorta Yorta tribe, and the snake-necked turtle was his totem.

As he spoke, I noticed a snake-necked turtle paused beside the road. I turned back and stopped, planning to carry it across the bitumen to safety. Looking down, I saw that the turtle was dead. A car tyre had shattered its brown shell. Meat ants hovered, tearing rotted flesh. Jimmy began singing a gentle song from his new album:

> Surely God was a lover
> When he bade the day begin
> Soft as a woman's eyelid
> Fine as a woman's skin
> Surely God was a lover
> All burning with desire
> When He called the night to come down
> And set the day on fire.[64]

Wallendbeen cemetery lies at the southern edge of Wallendbeen village, south from where I found the dead turtle as Jimmy Little sang. Inside the cemetery, apple box trees shade marble headstones and kangaroo grass.

In 1884, the Department of Lands paid the trustees of Wallendbeen cemetery seventy-five pounds to enclose three hectares beside the railway line.[65] But Wallendbeen village didn't grow into a town as government surveyors expected. Gravediggers left much of the reserved land alone. Behind a sturdy fence, undisturbed by livestock and ploughs, an array of grassy woodland plants continued to flourish across the fertile slope of deep red earth.

On paths and alongside the graves at Wallendbeen cemetery lie scattered ceramic shards, broken seashells and faded synthetic flowers. Headstones face east to receive the rising sun, a Christian symbol of resurrection and eternal life beyond the earth.

One memorial marks the grave of Samuel Hollis, who died in 1893 aged fifty-nine. 'Far Beyond This World of Changes', his epitaph reads:

> Far Beyond This World of Care,
> We Shall Find Our Missing Loved Ones
> In Our Father's Mansion Fair.

Another headstone memorialises William Palmer, buried in June 1916. According to his inscription on granite, heaven above is 'blessed', an 'everlasting home of peace and love.' By implication, earth is cursed, a transient place of violence and hatred.

Christians of recent centuries considered this world not a nourishing domain, explains environmental historian Roderick Nash, 'but a kind of halfway house of trial and testing from which one was released at death.'[66] Christians locate the source of all holiness in an eternal and unchanging heavenly home, a sacred domain beyond the dynamic, shifting nature of earthly life and embodied human existence. Death granted William Palmer transcendence from hostile terrain, his gravestone tells at Wallendbeen:

> At last when earth's days work is ended
> All meet thee in that blessed home above
> From whence thou camest where thou hast ascended
> Thy everlasting home of peace and love.

Banishing the sacred from earthly places dissolved moral constraints over human use of land and other species. Indeed, it was the moral duty of Christians to rearrange and simplify local ecologies. When Eve showed Adam how to pick fruit from the Tree of the Knowledge of Good and Evil, so the biblical story goes, God expelled the couple from the bountiful Garden of Eden into the desert. According to the story of Adam and Eve – the dominant religious narrative of western culture – intense agricultural labour, performed by men, recreates an inferior version of the lost Garden of Eden.[67] Christian teachings brim with agricultural instructions and metaphors. 'He that tilleth his land shall have plenty of bread: but he that followeth after vain persons shall have poverty enough', one proverb declares in the King James version of the Bible. Christ is the sacrificial Lamb of God, wheaten bread symbolic of his flesh.

In the 1880s, Australian inventor HV McKay developed the Sunshine Harvester, a labour-saving device able to strip, thresh and winnow wheat in one operation.[68] McKay apparently named the popular invention 'Sunshine' after attending a Christian sermon in Ballarat, west of Melbourne. The religion of God 'was all sunshine', McKay learned from the preacher, 'and the only difference between earth and heaven was that the sunshine of earth sometimes got overclouded, whilst that of heaven's was everlasting.'[69]

The new harvesting machine promised to bring the sunshine of God to the wheat paddocks and dark bushland of Australia. A device to make farming easier, the Sunshine Harvester helped settlers recreate the lost Garden of Eden across inland slopes and plains.

Medieval historian Lynn White famously described western Christianity as 'the most anthropocentric religion the world has seen.'[70] For nearly two thousand years, Christian missionaries cut down sacred groves in which others saw 'spirit in nature', White noted in his influential analysis of Genesis first published in 1967. Christianity established a dangerous dualism between 'man and nature'. Christians, argued White, imagined themselves as divorced from and superior to nature, a domain over which God willed them to rule.

Many theologians have challenged the arguments presented by White. Chapters in the Christian Bible other than Genesis, some have argued, encourage careful 'stewardship' of all the plants and animals created by God.[71] Despite the possibilities of alternative readings, however, the most influential interpretations of Christian texts in recent centuries cast humanity in firm control over the rest of nature.

Genesis informed the work of seventeenth century philosopher John Locke, an influential apologist for British colonisation. According to Locke, colonists following the biblical order to 'subdue the Earth' held moral and legal rights to terrain they secured and farmed. 'He that in Obedience to this Command of God,' wrote Locke,

> subdued, tilled and sowed any part of it, thereby annexed to it something that was his *Property*, which another had no Title to, nor could without injury take from him.[72]

In Australia, a continent without characteristically English arrangements of farmland and villages, Christian ideas helped justify colonisation and Aboriginal dispossession. Aborigines never cultivated land, colonists asserted, and therefore didn't own it. In the middle of the nineteenth century, the same paradigm lent righteousness and religiosity to the push for agricultural development and closer settlement across the inland pastoral estates of squatters.

On the southwest slopes my family engaged in a somewhat half-hearted way with Christian rituals and understandings. One summer more than three decades ago, we drove past wheat paddocks recently harvested into Junee and parked outside the Anglican church. In the red brick building, my sister and parents watched a priest baptise me into the Christian faith.

I returned to the church recently, carrying my certificate of baptism. 'We print the Cross upon thee here', the baptism card declares, 'And stamp thee His alone'. The wording of the certificate set me against earthbound forces and elements, instructing me 'manfully to fight' for Christ 'against sin, the world, and the devil'.

Inside the church, my hand felt the veined marble of the baptismal font, pale and cold. Slender windows – stained glass and pointed Gothic arches – drew my attention skyward, away from where I stood. Sunlight illuminated windows built into the northern wall. One glowing display of coloured glass showed a woman holding a golden sheaf of ripe wheat. Undulating farmland cloaked in crop extends behind her. 'In Loving Memory of Ethel Jane Ings', a brass plaque reads below the window, 'Faithful Servant of Christ and his Church 1888–1981.'

The hoot of a freight train rolling into Junee entered the church. I walked outside into warm autumn sun. Bees hovered, collecting pollen from a pink flowering gum. From the hillside footpath I gazed towards farmland beyond the domestic gardens and corrugated iron rooftops of Junee. One cultivated paddock of red earth lay in fallow. When autumn rains came, I imagined, a farmer would sow wheat or canola there.

Lining the street, elderly kurrajong trees threw shade over cars parked in front of the church. Across the southwest slopes the previous spring, kurrajongs had carried rampant sprays of creamy flowers streaked red. Now, tree branches held clumps of brown pods heavy with seed.

My grandfather was born inside Retreat homestead, north of Junee beside Pinchgut Creek, on Christmas Day in 1909. To mark the birth of their first son, his parents planted a kurrajong tree in the Retreat garden. My grandfather suffered various illnesses throughout his life. I remember the syringes he used each day to inject insulin into his thigh. At times when he became particularly ill, the growing kurrajong sickened too. His family watched the tree in the homestead garden, hoping that it and my grandfather would stay healthy and strong, imagining he and the kurrajong linked in some way. Today the tree is mature and flourishing, a robust memorial.

I picked some kurrajong seedpods before returning to my car. Roasted kurrajong seeds are tasty and sweet. Decades ago inside the Junee church, a priest asked a distant, heavenly god to grant me 'power and strength, to have victory, and to triumph against the devil, the world, and the flesh.'[73] Enmeshed in an industrial society of western and Christian traditions, can I refuse the power I didn't choose? As I drove away into late afternoon light, I pictured my grandfather, red earth and a sacred, nourishing grove of kurrajong trees.

silence

Unremembered voices

Retreat homestead has a deep verandah. Scratches across the timber floor – an expanse of white cypress pine felled in a nearby paddock – record generations of activity. A door leads into the homestead office at the western end of the shady verandah. I pause before inserting the key. Perfume rises from beds of violets and jonquils into unusually warm winter air. Beyond the garden fence, sunlight fills a paddock graced by elderly yellow box trees.

I sit on an oaken chair inside the dark and musty room, facing the roll-top desk. A framed photograph of stud rams – curled horns and heavy, rolling fleeces – hangs on the wall beside the desk. 'Kadlunga Stud Merinos, Sept 1912, two of the leading sires', the caption reads. Beneath red-brown dust, agricultural guidebooks and station records stand on shelves. I read a letter sent to my grandfather in 1956 by the Water Conservation and Irrigation Commission, granting approval for the excavation of a dam on Pinchgut Creek. 'The licensees shall destroy and keep destroyed within the whole of the area covered by the water stored by the dam all aquatic plants', the letter insists.

Chrome-plated veterinary equipment sits near an old scrapbook. Pasted inside, a newspaper cutting from November 1924 announces 'Another Sydney Record':

> At Tuesday's wool sales the Co-operative Wool and Produce Co. Ltd. included in its catalogue the well-known Retreat wool, on

account of Messrs. G. and H. Main, Retreat, Illabo. The top line of comeback, consisting of 5 bales marked SCB, realised 40½ d, which is a record for the Sydney market. This wool was a stylish and well-grown lot, well nourished, and of even quality throughout.

I open a yellowed booklet issued by the Graziers' Association of New South Wales, *1933 Award for shearing, crutching and wool-scouring operations and for station hands*. As the world economy faltered, the Commonwealth Court of Conciliation and Arbitration abandoned wage protection for Aboriginal pastoral workers:

> The definition of 'Station Hand' under the 1932 Award has been altered to exclude aborigines, who are consequently not entitled to receive Award wages. Their engagement is, therefore, a matter of mutual agreement and subject only to any special requirements of State legislation, such as the Aborigines' Protection Acts.

Inside the scrapbook, a second newspaper cutting catches my attention. The article from a 1926 edition of a South Australian newspaper begins:

> Mr F. H. Weston, well known as manager for many years of Kadlunga Sheep Stud, Mintaro, has returned from a visit to New South Wales, and on Friday he gave some interesting impressions of his trip.

Visiting the southwest slopes, Fred Weston, my grandfather's grandfather, stayed here at Retreat, west of Cootamundra, the home of his daughter and her family. 'Originally it was rather thickly timbered with box gum and pine', the journalist wrote of Retreat after speaking with Weston. 'It has been mostly cleared, but there is still a good deal of good pine and some gum timber left.'

During his stay in the district, the respected South Australian sheep breeder visited Widgeon Gully, a grazing station near Coolac:

> Widgeon Gully is a beautiful property of some 10 000 acres, with

a wide frontage to the Murrumbidgee. The river flats have a rich alluvial soil well suited to lucerne. Although only planted two years ago a wonderfully good cut had just been baled and carted. Three weeks later the plant was fully 12 in. high, and promised to give a good second cut, not withstanding that there has been no rain there for a considerable time. It is exceptionally dry. On one portion of the run, where two wethers to the acre have been running for upwards of 12 months, there is still an abundance of feed, and the wethers prime fat.

The newspaper published a grainy photo alongside the article. In his mid-seventies, Fred Weston looks old. He stands beside an Aboriginal man, roughly the same age. 'Mr. F. H. Weston and 'Marvellous', one of the last of the Murrumbidgee natives, who goes from station to station', the caption reads.

John Noble, nicknamed Marvellous, was a Wiradjuri man widely known across the Riverina and southern tablelands. Margaret Tucker, born beside the Murrumbidgee River on Warangesda mission near Darlington Point, remembered her grandmother's brother, 'Grandfather Noble', with great warmth. His nickname, Tucker explained, came from his habit of saying 'Ain't that marvellous?' at the end of each sentence.[1]

He was renowned for undertaking long journeys. Bob Glanville suspects that Noble was somehow related to his grandmother, Melinda McGuiness. When Bob's mother was a young girl, it always amazed her how the old man arrived in a clean white shirt after travelling swiftly on foot from afar. Noble would appear at the Cootamundra show to throw boomerangs and entertain the crowd, Bob told me.

When John Noble became sick and died at Cootamundra in 1928, newspapers speculated on his age. According to one newspaper, the elder was more than a century old.[2] Others said eighty or more.[3]

Noble was probably born on Muttama, a squatting run held in the early decades of pastoralism by Irish migrant Francis Taaffe.[4]

James Gormly, an early Riverina colonist whose father was an old friend of the Muttama squatter, thought Taaffe acted with justice and humanity towards Wiradjuri people. The squatter, Gormly noted, gave generous amounts of food to local clanspeople camped on the station.[5] Historical records suggest that violent encounters took place on neighbouring squatting runs upstream from Muttama, in the Cootamundra and Stockinbingal districts, around the time Noble was born. Taaffe may have offered rare protection and refuge to Noble's mother and her people.

On the southwest slopes, after an initial period of conflict and accommodation, Wiradjuri survivors lived relatively independently of squatters. During his youth and for much of his adulthood, Noble probably took part in ceremonies and fulfilled cultural obligations towards particular places and other species.

In the final decades of the nineteenth century, the extension of the Great Southern Railway and the ensuing tide of agricultural settlement transformed the region. As towns grew and diverse communities of grassy woodland plants and animals vanished, Wiradjuri could no longer maintain customary lifeways. Mary Gilmore described a final gathering of three hundred Wiradjuri northwest of Stockinbingal on the Bland Creek in about 1879, an event John Noble may have attended:

> It was only half the number expected, as those from the Lachlan could not get across in time, owing to drought. It was held at Morangarell, the home of my father's cousins, the McGregors. There I saw the last Gundagai chief. The blacks had called the meeting of the localised tribes, expecting and knowing it would be the last they would ever hold.[6]

In 1916 legislative changes granted the Aborigines Protection Board powers to make decisions for Aboriginal children without parental consent or court order.[7] The board advised the matron of Cootamundra Aboriginal Girls Training Home, established four years earlier, to expect a sudden inflow. 'I cannot forget any detail of that moment, it stands out as though it was yesterday',

Margaret Tucker wrote of the time she was removed from her family and sent to Cootamundra. One weekday in 1917, a policeman came to the school on Moonahculla mission, near Deniliquin southwest of Wagga, to collect Margaret, her sister May and another girl. Mission residents gathered outside the schoolroom door, Tucker remembered,

> all talking at once, some in the language, some in English, but all with a hopelessness, knowing they would not have the last say. Some looked angry, others had tears running down their cheeks.[8]

The schoolmistress, hoping to stop the removal, sent two boys to alert Theresa Clements, mother of Margaret and May, who worked at a nearby homestead. Theresa ran to the mission school without pausing to remove her apron. The policeman threatened to use handcuffs when she insisted her children stay. Theresa accompanied the girls to Deniliquin. Inside the police station, begging for the decision to be reversed, Theresa Clements heard the car engine start and rushed out.

'My last memory of her for many years was her waving pathetically', wrote Margaret Tucker, 'as we waved back and called out goodbye to her, but we were too far away for her to hear us.' Relatives found Theresa the next day, collapsed under a roadside tree, distraught.

Margaret Tucker remembered John Noble's great affection for her mother and the mission children. They reciprocated his love:

> My memory of this lovable old man was his kindness to us children and my mother, of whom he was very fond. He was my grandmother Bedgie's brother. Some children in those days felt he was a Witch Doctor. Mother and we children loved the old man, because he was good, although a bit cunning. He was very generous and would share his food with anyone. He loved his booze, and my father would scold him about this.

Between pages in the Retreat scrapbook, inside the homestead office, I find a photo fixed to a card – the image published alongside the newspaper report of Fred Weston's trip to the southwest slopes. 'Taken at Widgeon Gully N.S.W.', someone has written on the card in black ink. I lift the picture into daylight spilling through the office doorway.

The two grey-haired men stand closely, Weston slightly behind. Perhaps the photographer captured the image inside the Widgeon Gully garden. Dappled shade falls across the shoulders of the elderly pair. Noble, the shorter of the two, is dressed informally – his trousers creased, shirtsleeves rolled and collar open. Fred Weston is neatly attired and somewhat rotund, his right hand loosely supporting a pipe. The visiting sheep breeder appears content, settled. John Noble's eyes are sharper, his body lean, lips pressed. The Wiradjuri elder holds his arms close to his body, and clenches his fists.

Fred Weston and John Noble,
Widgeon Gully, Coolac district, 1926.

~ ~ ~

Beside rows of headstones at Wallendbeen cemetery, across the reserved hectares left alone by gravediggers, spring warmth evokes a colourful and sensuous display. Yellow blossoms of buttercups and yam daisies rise amid tussocks of kangaroo and snow grass turned grey by winter frosts. Purple, glossy flowers of chocolate lilies exude a delicious aroma.

I drove into the cemetery early one spring. Drawn by the new season, fresh green blades emerged through weathered tufts of kangaroo grass. Beneath red gum and apple box trees, hundreds of billy button flowers – each a single golden globe on a tall stem – hovered above the grassy surface like a miniature, earthbound galaxy.

Botanists have recorded more than fifty indigenous plant species inside the cemetery. Fenced to exclude sheep and cattle since 1884, Wallendbeen cemetery is a rare island of biological diversity and ecological connectivity. Nearby, yellow box trees sicken on hillsides blanketed in canola and wheat.

'If you stay in a white man's old burial ground long enough', writes storyteller William Least Heat-Moon, 'this darkness must come to you: his way of life is the land's death and his way of death is the land's life.'[9]

In 1968 anthropologist WEH Stanner called the denial by settler Australians of violent frontier encounters between Aborigines and colonists 'the Great Australian Silence'.[10] Tom Griffiths suggests that the silence was often 'white noise', historical narratives 'obscuring and overlaying' memories of atrocities. Denial of violence 'was frequently unconscious, or only half-conscious', writes Griffiths, 'for it was part of a genuine attempt by white Australians to foster emotional possession of the land'.[11]

Likewise, settlers on the southwest slopes constructed loud narratives to fill empty silences imposed by ecological erasure, and to make claims over land. 'The voice with which nature speaks is tactile, sensual, auditory, odoriferous, and visual', writes environmental historian Carolyn Merchant, 'a visceral understanding communicated through our hearts into our minds.'[12] Growing organisms express a lively genetic inheritance, a vibrant 'will to flourish'.[13] To replace biological diverse grassy woodlands and swamps with simple patchworks of crops and 'improved' pastures, colonists had to strenuously deny the will and expressions of native species.

Beside a busy intersection downhill from Wallendbeen cemetery are sculptures honouring the economic and productive success of the local wheat industry, with their fibre cables shifting green and yellow light through acrylic panels. Inspiration for the work, the brass plaque in Cootamundra explains, 'came from the wonderful visual effects of the wind rippling across vast expanses of wheat'.

The public artwork enshrines the modern, industrial model of broadacre monocultural wheat production. Production and profit are identified as primary local concerns. Attention is drawn to the food, industry and export dollars generated here.

Only simple, comfortable messages are offered by the sculptures on the outskirts of Wallendbeen village. Social and ecological burdens imposed by the development of industrial production systems are ignored. No references are made to dying eucalypts in nearby paddocks, to Wiradjuri dispossession, to the displacement of rural families by mechanisation and global competition. Events of the past are not linked to the present. Attention is directed towards a promising future of production and economic activity, and away from obligations to history and place. Processes of silencing are justified by the public artwork, the dominant industrial order bolstered. Rising above surrounding parkland, the sculptures stand like trophies erected by the winners of a scramble for economic security and profit.

In her poem 'Primeval Australia', Mary Gilmore called for remembrance of 'All that has vanished with the far-off summers.'[14] Violence had silenced 'multitudinous voices chirping to the sky'.

The poet described the displacement of the emu, 'a king of birds', by 'spreading herds' and 'great cities'. Gilmore pleaded,

> O Memory, tell of him to later comers,
> Lest he go unremembered with his summers!

Processes of 'unremembering' involve deliberate efforts to forget. Did Gilmore observe settlers turning away from the fragmentation and wounding imposed by forces of colonisation? Voices of the recently dead, denied and forgotten, could not easily rise to disturb and make claims on the present.

Gilmore did not seek comfortable solitude:

> Once all the whole year through the happy bush was loud;
> And, O the singing and the chatter after rain!
> Now on the plains the grass is like an empty shroud;
> The woods are silent, for the hand of man has slain:
> And you will never know, you later comers,
> What we, who pass, knew in the old far summers!

Powerful beliefs brought silences to rural Australia. 'A weed is a plant growing where it is not wanted', explain the authors of *Weeds: an illustrated botanical guide to the weeds of Australia*, published by the New South Wales Department of Agriculture in 1987. The representation of plants native to rural places as useless and dangerous justified the establishment of crop monocultures and simple pastures of several introduced species.

Agricultural scientists categorised as obstacles to production a range of plants Aborigines and some early settlers in southeast Australia found useful. The weeds guidebook published by the Department of Agriculture classes as problematic many species from which Aborigines harvested food and fibre, including cumbungi (*Thypha* spp.), warrigal greens (*Tetragonia tetragoniodes*), flax lily (*Dianella* spp.), bulbine lily (*Bulbine bulbosa*), tall spike rush (*Eleocharis sphacelata*), nardoo (*Marsilea drummondii*) and kangaroo apples (*Solanum aviculare*).¹⁵ On the southwest slopes, the development of export-oriented agriculture required the displacement of Wiradjuri people and the erasure of indigenous communities of plants and animals. Imperatives of empire, nation and the global marketplace determined what plants were classed as weeds. Strategies to deliver standard products to distant markets depended on emotional and cultural distance from local species and Wiradjuri people. Intimacy threatened the colonial project to harness the land.

Joseph Maiden, curator of the Technological Museum of New South Wales in Sydney, published *The useful native plants of Australia* in 1889. He described valuable attributes of many species disappearing across inland regions as pastoralism and farming intensified. 'Native Hops' harvested from hopbushes made 'beer of excellent quality'. The blossom of golden wattle, now Australia's floral emblem, offered a sweet perfume, Maiden noted, a sample of which was exhibited at the 1886 Colonial and Indian Exhibition in London. And golden wattle bark was one of the richest sources of tannin known in the world. Aborigines used the strong, silky fibres of flax lilies to make baskets. The reddish, mottled timber of drooping she-oak, Joseph Maiden explained, made lovely furniture. As burning wood, she-oak was unsurpassed. Weeping myall, also known as boree, a graceful small tree with pendulous branches, grew heavy timber from which Aborigines made boomerangs. Unpolished, the dark wood exuded a scent of violets. European craft workers made glove and handkerchief boxes from the fragrant timber, Maiden recorded.

When colonists arrived on the southwest slopes, a forest of boree cloaked parts of the Stockinbingal district, beside Bland Creek. Markets demanded livestock more than boree timber. By 1879, the Stockinbingal boree woodland was gone, a travelling reporter observed:

> From the remains of dead trees one sees how extensive it must once have been, but now not a single live specimen can be found. The cattle have eaten off the tender plants as they appeared above the

ground. The old ones have died through age, and the result has been the total extinction of the boree scrub.[16]

Later in life, Maiden seemingly came to value Australian plants for reasons beyond economics and utility. Settlers held ethical obligations, he believed, to protect native vegetation. 'To those people who put immediate utilitarianism in front of everything', wrote Maiden in 1908 when he was Director of Sydney Botanic Gardens,

> let me remind them that we are pioneers of a continent – a continent which possesses a remarkable and in many respects a unique vegetation; and the pleas of those who ask that the specialised vegetation should not be destroyed unnecessarily is worthy of some regard.[17]

He described a particularly large and 'beautifully rounded' Cootamundra wattle in the Bethungra district. The tree looked like 'a gigantic mushroom', sheep had nibbled every leaf from lower branches. 'Of course this was an exceptional tree', wrote Maiden,

> and it gives rise to painful reflections as to the wickedness of human nature, when one is informed that some miscreant has barked this tree, which was, without doubt, one of the most beautiful trees in the Colony.[18]

Northeast of Bethungra, at Cootamundra, devaluation of indigenous life brought profound silences. Before colonisation, Cootamundra was a place where diverse voices found expression. Gudhamangdhuray clanspeople allowed turtle populations to flourish unhampered in their swampland reserve on Muttama Creek. Autumn and winter rains soaking the stony ranges above Cootamundra fed hillside springs and the lively turtle sanctuary below. 'Cootamundra', a visiting journalist learned in 1896, was

> an aboriginal name signifying a marsh, the site of the town having been a lake or swamp, and used to be a series of gilgais and crab-holes in the lower parts, which are now drained by the growth of the town.[19]

Silence had descended over Cootamundra since the town was

established decades previously. The journalist offered a strident narrative of progress to expunge the quiet:

> Few, indeed, of the travellers who happened to camp near the site of the present town some thirty years ago only, and who were wont to be lulled to sleep by the sibilant sounds of insect life and the nocturnal croakings of the festive bull-frog, issuing from the swamp close by, could dream that the evolution of time has brought about such a change as has taken place here. The lake is dried up, the voice of the turtle and the quack of the wild duck is no longer heard in the land, the bull-frog is silent, and on the scene of these midnight revels has sprung up with remarkable rapidity a splendid town, of which more anon.

Parts of the turtle swamp survived into the twentieth century. Cootamundra barber Alan Crowe told me about a place below the high school, an undeveloped remnant of swampland locals called Frogs' Hollow. Alan remembers riding past at night to visit his wife and newborn baby at the Mercy Hospital. Many frogs called from the watery place, an area now covered by a football oval. Flat paddocks on either side of the Temora road just outside Cootamundra give a sense of the drained swamp. Here, after particularly heavy downpours upstream, earthworks can't contain Muttama Creek. Elderly eucalypts and tussocks of grass rise from a sheet of water.

According to anthropologist Alfred Howitt, Gudhamangdhuray clan territory centred on the swampy place where the town of Cootamundra grew. Members of the major southern Wiradjuri group defined themselves in relation to the swamp, the place where turtles were protected from hunting. Survivors of disease and violence maintained links with Cootamundra throughout the nineteenth century. In his writings James Gormly mentioned 'the blackfellows camp, which stood near Mr. Hurley's Cootamundra station' in the 1850s.[20]

When the Great Southern Railway arrived several decades later, agricultural development and closer settlement displaced Wiradjuri people from pastoral station campsites. Ecological fragmentation and the local extinction of many food species made life a struggle. Sanctuary law protecting turtles in Cootamundra swampland and other species elsewhere could no longer be enforced. Dispossessed and hungry families gathered at a fringe camp on the outskirts of Cootamundra.[21]

'Arranged with police to help me get the children tomorrow', Warangesda missionary John Gribble noted in his diary after arriving in Cootamundra in January 1882.[22] The missionary took fourteen 'mostly young Natives' to Warangesda, the *Cootamundra Herald* reported, far southwest beside the Murrumbidgee River.[23] In the Cootamundra camp, Gribble faced strong opposition from elders:

> Feb. 1st Rose early. Albert and I sought and found out blacks camp. Found about 30 men, women, and children, all in a sad state of semi-nakedness and hunger. Gave a man some money to buy bread. Talked kindly to all about Warangesda. Several seemed willing to go. But some of the older ones were very free in opposing my suggestion. I hope to get about a dozen away with me.

Perhaps with help from accompanying policemen, Gribble overcame the resistance of parents and elders:

> Friday 3rd. Took 12 poor waifs and strays from Cootamundra to the mission station. All at home gave the newcomers a most hearty welcome.

As Gribble took Wiradjuri children from their families at Cootamundra, he participated in the erasure of a rich cultural heritage. Oral cultures depend on the transmission of knowledge and language between generations. Understandings are contextual, rising from engagement with the particular natures of places. Far away at Warangesda, the newcomers couldn't hope to maintain knowledge of the intricate natural patterns and living systems of the Cootamundra district.

When the Aborigines Protection Board opened the Aboriginal Girls Training Home in 1912, Cootamundra too became a place where stories of country were silenced and lost. On a hilltop east of town, inside the former Cootamundra hospital, board staff trained Aboriginal girls from across New South Wales as domestic servants. The fringe camp at Cootamundra dispersed when the home opened. 'Aboriginal families steered well clear of Cootamundra over most of the 20th century', explains oral historian Peter Kabaila, 'principally to avoid any threat of child removal.'[24] Gudhamangdhuray clanspeople abandoned what remained of the turtle sanctuary below the granite hills.

Across Wiradjuri country and much of the continent, people lost places granting life and identity, and places lost people granting ecological stability and wellbeing. Death and displacement broke ties of mutual care.

Anthropologist Pamela Lukin Watson describes totemic relationships Karuwali people of southwest Queensland held with particular bodies of water. Before lowering themselves to drink, Karuwali paused and spoke gently, with reverence, to watery places. Ceremonies maintained abundant water supplies and bolstered the fertility of aquatic plant and animal species. When pastoralists murdered and evicted Karuwali clanspeople, water bodies became 'orphaned', and 'were said to cry out in pain.'[25]

On the southwest slopes and southern tablelands of New South Wales, Mary Gilmore remembered her father refusing to camp beside 'The Dead Water' places:

> 'The Dead Water' was a spring, soak, or waterhole at which no black sat, because all the group in whose walk-about it had been, were killed out. The dead were away, but the dead still owned it. No strange black visited it, no tribe trespassed on it. Like our cemeteries it belonged to those who were gone. There were many such places spoken of in the early days. Later on, the word was blotted out in a surveyor's name or forgotten. But when in 1872 we drove from Wagga Wagga to Goulburn, and back again, more than once we made the horses stretch at evening, The Dead Water being the only place for the nightly camp.[26]

As Wiradjuri abandoned the fringe settlement at Cootamundra and campfire embers dimmed to black, did the turtle swamp begin to grieve? Muttama Creek flows south from Cootamundra and joins the Murrumbidgee River upstream from Gundagai. Gilmore described a forlorn Murrumbidgee, mourning the loss of Wiradjuri people from riverside places:

> The Murrumbidgee whispering at its banks cries,
> 'Where are they
> Whose thousand camp-fires drove the darkness of the night away?'
> Silent the camps, the tawny embers cold,
> No more to throw on night their scattered gold.[27]

I walked along Muttama Creek through Cootamundra one spring afternoon. Decades ago, council workers on earthmoving machines enlarged the channel of the waterway on the northwest edge of town to contain floodwaters and protect urban property. Here, above the excavated channel, lies a section of formerly swampy earth. Exposed soil dead and dry, black with millennia of decayed life.

When heavy rains drench Cootamundra, earthworks and stormwater drains guide gushing water into Muttama Creek, away from flat areas where suburban homes and public buildings stand. Walking beside the waterway near the town centre, I noticed a vigorous mat of water couch almost bridging the channel. Grass creeps deeper into the stream, trapping silt carried by the flowing water. Gently, Gudhamangdhuray swampland attempts to reform.

To ensure the swift flow of stormwater along Muttama Creek and beyond town, Cootamundra Shire Council poisons cumbungi and water couch growing in the channel. Only after especially heavy rains will Muttama Creek overflow and inundate Cootamundra streets and domestic gardens. Swampland remains suppressed.

Cultural processes erase memories of Cootamundra swampland, complementing storm drains and the poisoning of water plants. A promotional publication issued by the Cootamundra Development Corporation reads:

> Set in a prosperous valley, Cootamundra is a picturesque town which is well-known as the birthplace of Sir Donald Bradman and as the home of Cootamundra wattle.

Donald Bradman is a national cricketing hero. Indigenous to the local area, Cootamundra wattle has attractive grey foliage and abundant bright yellow flowers, and is grown in many Australian gardens. Bradman and Cootamundra wattle are constructed and presented as town icons because they are valued beyond Cootamundra, in the national and global arenas the southwest slopes are harnessed to serve. The promotional publication makes no references to the Wiradjuri origins and rich cultural context of the placename 'Cootamundra'. Modern cultural processes obscure and deny local elements – ecologies, species, histories – irrelevant to powerful systems of economics and global trade.

In the twentieth century, colonial processes suppressed resistance to the removal of Aboriginal children by officers of the Aborigines Protection Board. 'We Aborigines had no say', explained Iris Clayton, a former resident of Cootamundra Aboriginal Girls Training Home.[28] Clayton was born in 1945 at Leeton, a town in the Murrumbidgee Irrigation Area northwest of Narrandera. Her family lived alongside other Wiradjuri and poor settlers at Wattle Hill, a shantytown just outside Leeton. Huts of tin and timber salvaged from the nearby rubbish tip stood in rows across the hillside. Residents of the fringe settlement worked in the local cannery and on irrigation farms picking fruit.[29]

When Iris Clayton was eleven years old, Aborigines Protection Board officers visited Wattle Hill to speak with her mother:

> Mum had one day to pack our things and we never said goodbye to my grandmothers. I remember Mum scrubbing us all in our big round galvanised bath tub. She nearly took the skin off us. Then the taxi arrived early for the eight o'clock train. Aunty Lulla came running down the hill crying and pressed her last ten shilling note into my hand.[30]

James Morgan grew up at Wattle Hill. He remembered policemen and a welfare officer arriving to take eight children from one family. Official power, Morgan explained, stifled Aboriginal opposition to child removal:

There was a huge crying and screaming, because they knew they were being taken. They were trying to run away and hide and were being pulled out by their legs from under the beds. Their mother was there. That's why the police were there, to enforce orders given by the welfare officer. He stood by and organised it, but it was the cops' job. They took the whole lot, 8 girls and boys. Their mother stayed for a while. I don't know what became of her after that. People stood outside and watched, but wouldn't do anything because they were too frightened. In those days the police could just go into your house and do whatever they wanted. You couldn't go to anyone to complain, you couldn't say anything to anybody.[31]

At Cootamundra inside the Aboriginal Girls Training Home, Protection Board officers tried to expunge Aboriginal identity and culture from the hearts and minds of residents. 'Our traditional language was banned and punishment was meted out to those who used it', recalled Clayton.[32] Lesley Whitton, another former resident of the home, explained the agenda of the Aborigines Protection Board. Erasure of Aboriginality offered Aboriginal girls a position at the bottom of Australian society. Western notions of hierarchical social order prohibited alternative ways of being:

> At the Home we were taught to cook, sew, wash and iron. Some of the girls were obviously unhappy and unsettled. The police brought them back. I've never been involved in Aboriginal issues before, because we were not taught Aboriginal culture at the Home. Some of the girls who were reared in the culture on the missions and reserves lost out when they were moved to the Home. Aboriginal culture was cut right there.[33]

Like other nations bound to global networks of trade and power, the Australian state couldn't tolerate indigenous lifeways and connections to place. On the southwest slopes, European colonists now held Wiradjuri country. Harnessed for export-oriented production, the fertile region generated national wealth. Settlers identified and removed barriers to the pursuit of colonial and national objectives. Town councils and the Aborigines Protection

Board encouraged Wiradjuri to settle on reserves established outside the productive farming region – Warangesda on the Riverine Plain, Hollywood at Yass on the southern tablelands and Brungle in the foothills of the Snowy Mountains – where land was cheap and settlement less intense.[34] Displacement of Wiradjuri, and the erasure of language and culture, removed and silenced a group of people opposed to colonisation.

Wiradjuri elder Stan Grant grew up on an Aboriginal reserve beside the Lachlan River at Condobolin. 'If you don't have your language, you don't have a culture', he told me. Stan remembers elderly Wiradjuri making sure children didn't learn the tribal language. 'It was a real threat', Stan said,

> the fear was real, that had we gone to school and started speaking Wiradjuri, that the governments may've come and removed us and took us away from there. Because the threat was there, all the time: 'if you speak this language to your kids we'll come and remove your kids', you know. So the language was never spoken around us.

As the twentieth century drew to a close, Aboriginal activists and supporters brought radical change to community attitudes and government policy. Possibilities opened for Wiradjuri people to make firm political demands. In 1993, one year after the High Court of Australia recognised native title, a group of Wiradjuri, including lawyer Paul Coe, claimed ownership of the vast region once owned by their people. 'There was no legal or moral right to take the Wiradjuri land', Coe argued. 'We're seeking more than native title; we're seeking sovereignty.'[35]

The ambitious claim challenged dominant settler narratives of agricultural progress and Australian nationhood. NSW Farmers' Association representative Terry Ryan drew upon traditional settler stories of national development to oppose the Wiradjuri action. The area claimed 'included land regarded as some of the nation's finest agricultural and pastoral land, producing wheat, rice and other crops', Ryan said. Employing notions of inevitable linear progress and insisting on a firm divide between past and present, *The Land* newspaper likewise reflected popular opinion: 'we can't turn the clock back and nor should we try'.

> The British colonisation of Australia triggered the birth of a new nation. History is littered with cases of invading tribes and aggressive nations controlling and supplanting others. Non-aboriginal Australians of today weren't responsible for the decision to colonise Australia and they shouldn't feel guilty about the treatment of Aborigines by a minority of early settlers. Australia exists, history can't be rewritten.[36]

Challenged by the Wiradjuri claim, *The Land* and other respondents sought to reaffirm dominant versions of history and shield the powerful interests those narratives serve. Paul Coe presented a different, unsettling story of colonial processes. Settlers had 'trespassed upon the lands of the Wiradjuri nation and forcibly, wrongfully and unlawfully dispossessed' its people.

Despite the unlikelihood of the claim succeeding, the alternative historical narrative told by Coe induced virulent responses. Tim Fischer, National Party leader and a farmer from Boree Creek near Narrandera, warned of a 'violent backlash among the white community.'[37] Farmers had weapons, 'armed and ready', said one member of Wagga City Council.[38] According to John White, executive director of NSW Farmers, landholders would fight Wiradjuri people outside the legal system if the claim did succeed:

> Farmers will treat this very seriously. People are not going to walk off their land without a fight. This will lead to growing suspicions against Aboriginal people and hatred and even, in the end, violence.[39]

Sir Anthony Mason, Chief Justice of the High Court, deemed the action taken by Paul Coe and fellow Wiradjuri 'improper' and dismissed the claim.[40]

Five years later, encouraged by the newly formed local reconciliation group and Wiradjuri elder Pastor Cec Grant, Cootamundra Shire Council erected road signs identifying the Cootamundra district as part of the traditional lands of the Wiradjuri people. Like the Wiradjuri land claim made in 1993, the 'Wiradjuri Country' road signs on Cootamundra Shire boundaries suggested a traumatic history of dispossession and erasure, challenging dominant narratives of progress and agricultural development.

There was 'mixed reaction' in the local community to the new signs, the *Cootamundra Herald* reported, and shire councillors received 'a number of comments from concerned residents.'[41] Driving along the Harden road recently, I noticed someone had again stolen the timber 'Wiradjuri Country' sign at the Cootamundra Shire boundary. Since the sign went up in 1998, the council has replaced it several times.

Desire to silence disruptive voices remains present and active in the Cootamundra district. Rather than identifying possibilities for peaceful coexistence, it appears that the vandal and those locals concerned about the road signs are threatened by Wiradjuri and settler efforts at reconciliation.

Dialogue enables healing of ecological and social wounds. Silencing only furthers division and fragmentation. Fortunately, people are listening to painful stories and fostering change. In the late 1990s, as Wiradjuri and other Aboriginal people told alternative narratives of Australian colonisation, Cootamundra residents joined a hopeful national process of dialogue and healing. On the pages of a Sorry Book distributed by the Council for Aboriginal Reconciliation, now displayed inside the Cootamundra Heritage Centre, many locals recorded their sorrow over injustices committed by settlers: 'So very, very sorry', 'I am sorry for the part my extended family played in causing your pain', 'May your sorrow become our sorrow', 'I'm sorry for what you've all been through and the scars that you have had cut through your souls that will never heal', 'My love to you all'.

As the reconciliation movement at times demonstrates, settler narratives and ways of seeing destructive to Aboriginal people and to relations between settlers and Aborigines may be critiqued and rejected. Beliefs, attitudes and cultural processes underlying the disordering of local ecologies may likewise be challenged and dismantled. Alternative stories and realities are possible.

Red steers and white death

Good rain fell across the southwest slopes during the winter and spring of 1986. Bulky pastures and crops turned to gold as summer approached. January days were dry and hot. Under bright summer skies, I spent my school holidays on a farm north of Cootamundra, working through the harvest.

One hot and windy afternoon, we watched a great plume of smoke – tall and bulbous like an atomic mushroom cloud – rise in the south. On a hillside north of Junee, a man cutting thistles and burrs had struck a stone with a hoe and sparked a catastrophe. Dry and gusty westerly winds pushed flames through paddocks, across roads, beyond frantic firefighters. Next day, the forested range beside Bethungra was alight.

Winds subsided and temperatures fell as night came. In the darkness, workers burned back into the hills and graded wide swathes on lower slopes. But fears were realised the following day as temperatures climbed and erratic gusts lifted burning embers over containment lines. Three new fire fronts swept eastward with extraordinary intensity. Witnesses described great fireballs running across hills, walls of fire igniting paddocks a kilometre ahead, and flames taller than thirty metre pine trees. Fire truck drivers couldn't keep pace. One farmer recalled a tube of fire speeding across paddocks, flames coiled and rolling – the fabled 'red steer' of Australian wildfire.

Firefighting coordinators kept track of developments from a helicopter. Sydney bushfire brigades arrived in large trucks fitted with advanced equipment. To save homesteads, a local crop-spraying pilot dropped water over encroaching flames. Bethungra residents abandoned homes as the blaze approached. Police reported 'widespread panic and general chaos'. Likened to a 'steam train running on full power', the inferno pushed east towards Muttama.

In one tragic incident, a ball of flame enveloped firefighters Allan and Paul Rolles. Severely burned, the father and son drove back to their farm near Gundagai. Allan asked his wife Patsy to spray them with a garden hose. Treated in local hospitals then transferred by air ambulance to Sydney, both men died in the burns unit of Concord Hospital.

Cooler weather and light rain helped firefighters contain and eventually extinguish the inferno. In three days and across thirty farms, the fire had consumed more than twenty thousand hectares. Twenty-five thousand sheep and cattle perished. Farmers buried scorched, blistered bodies in deep pits, and wondered how to begin restoring their devastated properties.

Later, in a Muttama park, the local bushfire brigade built a quartz stone cairn in memory of Allan and Paul Rolles. It stands under a mature ribbon gum. When I saw the memorial one autumn, long strips of bark swayed from branches above the cairn. Some nearly touched the ground. Most varieties of eucalypts ache for the heat and power of flame. Crisp streamers of dry bark peel away and invite fire upward. Seedpods burst as flames build. Once fire passes, fresh leaves sprout from lignotubers. With rain, seed germinates on surfaces burned clear of grasses and shrubs.

Tragically, the shady area under the ribbon gum seemed an appropriate position for the memorial cairn of white stones. Over millions of years, fire has shaped and energised local ecologies across Australia. Efforts by settlers to suppress the natural force of fire continue to bring dreadful harm to land and people.

Mary Gilmore noted differences between settler and Wiradjuri responses to summer bushfires on the southwest slopes more than a century before the devastating events of January 1987.[42] 'I have seen a whole station in a panic', wrote Gilmore, men, women and children nearly killing themselves with frantic and wasteful effort; and then a handful of blacks and lubras under their chief come and have the fire confined and checked in no time. Having the confidence of habit they allowed the fire freedom where it seemed least dangerous. In one such fire they concentrated on the sides, letting the centre flame run forward. But far in advance of this ran lubras hunkering down over their half-yard-wide flares. Behind the first row a second line was at work, and behind this a third, each fire opposite the gaps between the forward ones. The advancing tongues of flame having been kept narrow by attention to the sides, the draught was narrow, so a very wide front of little fires was not necessary. When the advance met the little islands of burnt grass it died there; in the lanes between it was beaten out. The chief told my father that unless fire was kept narrow and beaten out before it created a high wind it was no use trying to fight it. Once it created its own such wind it was invincible.

Settlers applied force to contain fire outbreaks. In contrast, Wiradjuri relied on detailed understanding and subtle action:

There was a difference between the blacks' methods and the white's. The white man used large bushes and tired himself out with their weight and by heavy blows; the blacks took small bushes and used little and light action. The white expended the energy of panic; the blacks acted in familiarity, as knowing how and what to do. They used arm action only, where the white man used his whole body. Where, as a last resort, the white man lit a roaring and continuous fire-break, the aboriginal set the lubras to make tiny flares, each separate, each put out in turn, and all lit roughly in line. The beaters they used were so small that they hunkered to do the lighting and beating.

Wiradjuri and other Aboriginal peoples considered fire 'a major totem, a friend', writes historian Bill Gammage.[43]

Settlers did find uses for fire. Smoke haze cloaks the southwest slopes of New South Wales in autumn, as farmers burn the sunfaded stubble of crops grown the previous year. Nevertheless, most settlers came to see fire primarily as a natural force to be feared, not an amiable agency to befriend. Eucalypts, wattles, chocolate lilies, yams, cumbungi and other native plants flourish after the passage of fire. Not so annual crops and pastures turned dead and dry by summer heat. In the twentieth century, Australians developed sophisticated firefighting trucks, water-bombing aircraft and other powerful technologies to master bushfires and protect paddocks of imported plant and animal species.

'Men pay for the increase of their power with alienation from that over which they exercise their power', philosophers Theodor Adorno and Max Horkheimer observed in their famous critique of the Enlightenment.[44] Since the beginning of British colonisation, settler efforts to master fire and dynamic rural terrains have blocked close learning. Powerful relationships are monological. Understanding does not easily flow both ways. Quests to suppress fire have obscured the interdependency of fire and the living systems of Australia. Humanity is the 'fire agent' of the biosphere, argues fire historian Stephen Pyne. As such, people hold a 'duty of care to the living world' to recognise their 'ecological presence' and to nourish local ecologies with careful burning.[45]

Panic behaviour during bushfire, as observed by Gilmore and as reported by Cootamundra police in January 1987, is one example of fearful responses to powerful expressions of a natural force. Similarly, Australians show fear in the face of farmland salinisation. In the late 1990s, scientific forecasts of escalating salinisation destroying infrastructure and limiting agricultural production captured public attention. People imagined the salinisation process as a frightening beast stalking inland paddocks. Journalists and politicians spoke fearfully of a 'silent, creeping menace', a force 'that threatens our farmers', the 'dryland killer choking the land's lifeblood', an 'insidious poison', 'a cancer slowly creeping along the system', 'the White Death'.[46]

Are such descriptions more than the hyperbole of political speechwriters and reporters? Perhaps a haunting and overdue realisation underlies these vivid responses to dryland salinisation.

Rising, salty watertables draw attention to the ecological limits of farmland. Salinity challenges the established, demanding models of agricultural science and modern farming. In reaction to the industrial project of transcending natural realities, the land imposes new constraints.

As scientists reveal the seriousness of ecological disorders in agricultural regions, a widespread lack of familiarity with natural patterns evokes fearful responses. Few Australians, it seems, understand how modern agricultural systems induce dryland salinisation and other ecological problems. In turn, few can imagine how to address those disorders presently undermining food and natural fibre production. Kentucky farmer Wendell Berry observes that

> when we have destroyed the forests and prairies to replace them with agriculture we have never known what we were doing because we have never known what we were *un*doing.[47]

A similar process took place on the western slopes of New South Wales. British settlers imported ancient cultural traditions of grazing and farming, faith in modern industrial strategies, and a restless desire for change. Ignorant of local ecologies, colonists erased grassy woodland and built patchworks of crops and pasture.

Rural landscape images of country homesteads, wire fences, old paddock trees, golden crops and iron woolsheds are familiar to most Australians. Complex links between ecological disorder and the operation of industrial farmland are rarely understood.

Cultural analyst George Myerson observed a 'dark relegitimation of the modern order of things' at work in Britain as people responded to global warming, 'mad cow disease' and other phenomena.[48] Faith in narratives of scientific and technological progress faltered as the disorders arose, until science and industry generated explanations and remedies. In each case, the modern promise of knowledgeable control was eventually restored and a widespread crisis in confidence averted. 'Industrialism always proposes to correct its errors and excesses by more industrialisation', notes Wendell Berry.[49]

In the 1990s, scientists and technologists tried to explain and arrest the spectre of dryland salinisation, a process induced by the vigorous application of abstract, universal scientific theories and industrial technology. 'Scientific and technological innovation both on farm and in laboratory will play a fundamental and increasing role in the development of sustainable farming', CSIRO Land and Water scientist John Williams argued recently.[50]

But scientific and technological solutions designed to fit within the dominant modern order inadvertently reinforce tendencies for neglectful and destructive actions. Any critique of underlying cultural and institutional dynamics is avoided. Imperial policies, economic constraints and imported frameworks of perception and belief drove colonists to destroy and displace Aboriginal clans, harness land for primary production and disrupt local ecologies. Historical and cultural processes shaping destructive activity and blocking dialogue between people and rural places, not the absence of sophisticated western scientific knowledge or technological capacity, are the primary causes of ecological disorder in agricultural regions.

Responses to dryland salinity often demonise the salinisation process itself. Attention is directed away from causes, from the industrial farming practices and globalised economic dynamics straining rural places. Responses that are focused on symptoms shield the dominant model of primary production and agricultural trade from critique. Dryland salinisation is cast as a new 'weed' or 'feral', an insidious constraint on production for agricultural scientists and technologists to overcome.

Technology and science meet salinisation with force. 'Scientists have made a breakthrough that will give Queensland farmers a simple but effective tool in their fight against salinity', a Queensland rural newspaper announced in 2002:

> Field research carried out in the Murray-Darling Basin is expected to lead to the development of new technology that will detect salinity before it surfaces, allowing both irrigation and dryland farmers to take preventative measures to slow its progress, and potentially even stop it developing.[51]

The old battle for mastery over the natural forces of the land is reinvigorated and relegitimised, the language of war reapplied. A recent article in the agribusiness section of *The Land* newspaper told readers:

> Australia's war on salinity has received a boost with the release of two new CD ROMs containing a wealth of information about natural resource management tools, models, frameworks and mapping programs.[52]

The Australian Academy of Science likewise uses militaristic rhetoric to explain the monitoring of salinisation with electromagnetic photography from planes and satellites:

> Armed with the information such methods will provide, a coordinated community response could succeed in combating the white death, before it eats out our agricultural heart.[53]

Emotive responses to dryland salinisation are often infused with nationalism. Dryland salinity is a 'creeping white tomb that is overwhelming farmland across the nation', a beast stalking 'our farmers', ready to consume the 'agricultural heart' of Australia.[54] Militant, co-ordinated efforts are needed to defeat enemies of the nation. The federal government has declared a national 'war against salinity'.[55] Looking beyond state borders, the Nature Conservation Council of New South Wales believes a national 'environment levy' on income tax is required to fund the 'fight' against salinisation.[56] Similarly, the Murray Darling Basin Commission recently called for a national 'environmental services levy' to raise more than sixty billion dollars over ten years. The massive amount could rehabilitate the fertile slopes and plains of inland southeast Australia, an area growing 'over 40 per cent ($12 billion) of the nation's agricultural production'.⁵⁷

Unless care is taken, national plans seeking to address problems imagined as national in scale may block possibilities for local actions and adjustments. Like the turn towards science and industry to explain and control dryland salinisation, calls to nationalist sympathies tend to reinforce other cultural and economic foundations of ecological disorder in agricultural regions. Nationalist sympathies framed in economic terms inadvertently encourage neglect towards components of rural places lacking immediate value in the global marketplace.

In the nineteenth century, imperial policies and the economic demands of distant markets drove the development of industrial farmland and transport routes across the southwest slopes. Few colonists knew what was needed for local ecologies to remain strong and naturally productive. Industrial systems of agricultural production driven by forces rising elsewhere blocked opportunities to sense and understand the ecological dynamics of rural places.

Dryland salinisation and other disorders in agricultural landscapes are products of entrenched tendencies to deny and fragment local components and patterns of connection in favour of national and global imperatives. Rather than seeking intimate ecological understanding and the integration of primary production into natural systems, settlers applied mechanical devices and universal scientific theories to erase and overcome particular, imagined constraints of rural places.

In the second half of the twentieth century, industrialisation and farm expansion forced thousands of farmers and workers away from agriculture and rural regions. Great machines propelled by fossil fuels banished humans from farmland. The process of agricultural industrialisation and rural depopulation, Wes Jackson argues, exemplifies 'a law of human ecology: high energy destroys information'.⁵⁸

Technological development, globalised economies of scale and government policies are today intensifying a situation whereby fewer people work larger farms. Departing farmers and rural workers take knowledge and understandings garnered over decades and generations of labour and observation. On wide properties emptied of people and memories, there is less potential for farm managers and workers to learn the intricacies and particularities of rural places. Responsive relationships between people and land are impeded. Without populous rural communities and intimate understandings of local ecologies, farming cannot be folded into the complex and shifting patterns of nature.

As the government and community-funded Landcare movement emerged in the 1990s, environmental scientists and farmers developed methods to restore and maintain the productive capacity of farmland. Landcare has fostered valuable and widespread learning about local ecologies. Unfortunately, the movement hasn't encouraged critique of the economic and cultural dynamics underlying ecological disorder. In farming today, production and profit maximisation remain the primary goals and standards of measure. Landcare reaffirmed the cultural tendencies responsible for disordering local ecologies. Farmland remained a 'natural resource' to be harnessed and managed with western science and technology for the good of humanity and the national economy. While the Landcare movement did foster dialogue between people and rural places, global economic forces and government policies favouring competition ensured monological relations of power over subjugated farmland remained the norm.

The Australian Broadcasting Corporation recently produced a major documentary series, *The silent flood*, examining the phenomena of land and water salinisation induced by settler activities. Series producers described the process of salinisation as 'the biggest environmental threat to Australia in the 21st century'.

The title 'silent flood' implies a perception of salinisation taking place quietly. Dryland salinity scalds appear quietly, plants die quietly, animals and insects vanish quietly. 'Silent flood' makes sense and is marketable nationwide because efforts at mastery continue to deafen Australians to the varied expressions of rural places.

Local ecologies, plants and animals offer messages intelligible to people. In the field of modern agriculture, non-human agencies and natural forces present in rural places are sidelined and silenced. Different understandings arise when people seek dialogue with farmland and other species. When we turn towards rural places harnessed within industrial systems of production, land and stricken beings may be heard to scream, to weep.

Defending agriculture

One night I stuck a poster, *Conserving grassy box woodlands*, to the kitchen wall inside our rented farm cottage near Wallendbeen. A photo in the middle of the poster shows a diverse spread of native grasses and flowering forbs under box trees, somewhere on the western slopes. Text describes the ecological interactivity of healthy grassy woodland remnants. A particular species of wasp, the poster explains, pollinates the purple donkey orchid. If the insect disappears, so does the orchid. Squirrel gliders inhabit tree hollows in daytime. The small possums with big ears and bushy tails emerge at night to eat insects, including pest species partly responsible for the premature dieback of paddock trees. 'Squirrel gliders are becoming very rare', the poster reads.

Inside the old weatherboard cottage, I imagined the hillside beyond the kitchen window – moonlight illuminating a ripening wheat crop and, where slopes meet below, a haunting tracery of dead yellow box trees. Wallendbeen farmland is renowned for carrying heavy crops, for generating staple foods and export dollars. In western agricultural societies, abundant production tends to equate with good farming and signify the moral worth and status of farmers.

George Sessions, a progenitor of the deep ecology movement, describes the subversive quality of an ecological sense of natural connectivity and interdependence.[59] I felt uncomfortable fixing the poster about grassy woodland conservation to the wall inside the Wallendbeen cottage. Rather than the sort of dynamic ecological relations described in the poster, modern systems of primary production rely on powerful, interventionist technologies. The poster seemed improper on display, a rejection of my culture and my people.

There is little space in rural Australia to value natural patterns devalued and erased by industrial agriculture. People draw identity, status and income from rural property ownership and the application of vigorous farming strategies. Notions of exclusive rights over rural places enable agribusiness activity and stifle critique. In his paper 'Land management and asset security', former National Farmers' Federation president Ian Donges argues for the recognition of absolute and exclusive rights over land, a set of rights extending only to the future generations of individual farming families:

> Farmers make the day-to-day management decisions affecting more than 70 per cent of Australia's land mass and 70 per cent of the water diverted from our rivers and streams. They are the most important stakeholders in good environmental management. They are passionate about protecting our natural resource base in a bid to ensure sustainable agricultural industries for their children and their children's children.[60]

Exclusive private property rights over farmland, a 'natural resource base' for industrial agriculture, are deeply respected in Australia. Farming and agribusiness representatives use the rhetoric of exclusive rights to resist the legitimate interests of other groups and individuals in rural places. Entrenched, dichotomised ways of knowing set rural and urban domains in opposition. Intimate ties that every town and city dweller has with farmland are denied and obscured.

We are all nourished, our bodies kept warm and alive, by the scattered rural places where food and natural fibres are grown. Agriculture connects everyone 'in the most vital, constant, and concrete way to the natural world', writes environmental historian Donald Worster.[61] Reciprocating the life and wellbeing offered by farmland, caring for the living systems in which we are all embedded, has never depended on private ownership. 'On the contrary', Worster observes, 'possession has often led to alienation of affection, to exploitation and indifference.'[62] Exclusive claims over rural places turn people away, leaving scientists and farmers to address ecological disorders alone.

In western societies, scientists are often asked to provide all necessary knowledge and find all necessary solutions. Such faith in the power of science, Wendell Berry notes, reflects 'a general abdication of our responsibility to be critical and, above all, self-critical.'[63] Widespread questioning of modern systems of primary production, processing, marketing and consumption is needed. Ecological scientists reveal escalating disorder in Australian agricultural regions. An adequate response requires acceptance by urban as well as rural dwellers of ethical obligations to care for life-giving rural places.

Modern agriculture is a vulnerable activity. Landholders depend on the productive success of one or several genetically uniform species contained inside paddocks. In the early decades of British settlement on the southwest slopes, Wiradjuri sometimes attacked and destroyed entire cattle herds, often the primary income source for squatters. Today, the Commonwealth Department of Agriculture, Fisheries and Forestry operates an 'Office of the Chief Plant Protection Officer', and NSW Agriculture employs 'Agricultural Protection Officers'.

Mirroring the security concerns of the Australian state, Plant Health Australia, funded by the federal government, adopted anti-terrorism rhetoric in a recent newspaper ad about the protection of crops from exotic diseases and insects. 'Spotted anything unusual?'

the advertisement asks. 'LOOK. BE ALERT. CALL AN EXPERT.' The ad gives a phone number for a free 'Exotic Plant Pest Hotline'.[64]

In 2003, scientists were alarmed to discover plants affected by wheat streak mosaic virus in CSIRO breeding trial sites. The Grains Council of Australia feared the disease would cut the value of the national wheat crop by 200 million dollars each year.[65] To contain and eradicate the virus, the federal government swiftly implemented an emergency management plan.[66]

In Australia, variable climate and weather patterns intensify the vulnerability of modern industrial farming. One night in October 2001, a heavy frost descended across the southwest slopes, a rare and dreaded event. Crops in spring flower are particularly vulnerable. Ice crystals turn flowers sterile, and plants grow no grain. NSW Farmers estimated that the late frosts cost primary producers on the southwest and central slopes 100 million dollars.[67] In some areas, frost ruined 80 per cent of crops.[68]

The organisation called for emergency federal funds to help suffering farmers.[69] 'This state of affairs shows just what a gamble farming is', said Illabo farmer David Carter after the disastrous cold snap. 'There is certainly no need to go to Las Vegas, farming is the biggest gamble of all.'[70]

The vulnerability of modern agriculture strengthens the sense of moral purpose infusing the culture of the industry. In Australia, models of farming imported from western Europe remain particularly vulnerable, despite modifications made in response to local conditions.

Erratic weather and climate patterns heighten the moral pedestal onto which farming is raised. Since the beginning of colonisation, settlers have honoured the hard work needed to raise imported varieties of animals and crops across unfamiliar and unpredictable terrains. Agricultural activity signified civilisation and colonial success.

In 1924, the *Agricultural Gazette of New South Wales* published a short article by John Strong, Professor of Education at Leeds University. 'History and literature and art have shown throughout the ages', wrote Strong, that daily contact with the elemental forces of nature breeds independence of character, virility of mind, constancy of purpose — qualities included among those accounted worth while in life.[71]

Cultivation of soil, many believed, reflected a cultivated and superior society. Agriculture distinguishes 'man' not only 'from the inferior animals around him', the same publication declared in 1898, 'but also indicates the difference between the savage and the civilised state of his own species.'[72]

In Australia, belief in the especial dignity of rural activities underlay a succession of closer settlement policies implemented across inland slopes and plains. Agricultural development, people thought, offered the new Australian nation a stable social and economic foundation.

Towards the end of the nineteenth century, swiftly rising European populations generated fears of food shortages. Australian farmers and agricultural scientists responded vigorously to western European demands for wheat.[73] Widespread application of new farming methods and technologies saw Australia become a major wheat exporter early in the twentieth century. Popular, nationalistic writers extolled Australian agriculture and rural life. A farmer character imagined by celebrated poet CJ Dennis knew the nobility of his profession:

> Wheat, Wheat, Wheat! When it comes my turn to meet
> Death the Reaper, an' the Keeper of the Judgment Book I greet,
> Then I'll face 'em sort o' calmer with the solace of the farmer
> That he's fed a million brothers with his Wheat, Wheat, Wheat.[74]

Nationalist poet Banjo Paterson likewise imbued wheat farming with moral integrity. 'Song of the Wheat' tells of agricultural development across former squatting runs on the southwest slopes and plains of New South Wales.[75] Paterson honoured indefatigable settlers erasing grassy woodlands to build farms:

> Yarran and Myall and Box and Pine —
> 'Twas axe and fire for all;

> They scarce could tarry to blaze the line
> Or wait for the trees to fall,
> Ere the team was yoked and the gates flung wide,
> And the dust of the horses' feet
> Rose up like a pillar of smoke to guide
> The wonderful march of Wheat.

In summertime, as wheat harvesting drew to a close, hard working farmers could rest with satisfaction, wrote Paterson:

> When the burning harvest sun sinks low,
> And shadows stretch on the plain,
> The roaring strippers come and go
> Like ships on a sea of grain,
> Till the lurching, groaning wagons bear
> Their tale of the load complete
> Of the world's great work he has done his share
> Who has garnered a crop of wheat.

Later in the twentieth century, as nations recovered from World War II, agricultural scientists and other proponents of industrial farming systems worried about global food security. 'The underdeveloped world is losing the capacity to feed itself – it is losing the race between production and reproduction', agricultural scientist Eric Underwood warned in 1967.[76]

Reasons for famine are varied and complex. When European powers began colonising distant lands centuries ago, interrelated processes of eviction and starvation transformed subjugated communities. Powerful colonial interests secured land from hunter-gatherers and subsistence farmers to produce expensive export commodities.

The process continues today under economic globalisation. People dispossessed of land must purchase food or go hungry. Food is abundant in the global marketplace. Access to food requires wealth. When world hunger is discussed, agricultural scientists and agribusiness representatives tend to ignore complex historical and economic contexts. Instead, they argue for the intensification and spread of industrial farming. 'As our communities burst at the seams with ever-increasing numbers of mouths to feed, the demands on the production base will grow', said Ian Macdonald, New South Wales Minister for Agriculture and Fisheries, in 2003.[77] The only option, Macdonald argued, was 'continual refinement of existing technologies – and the development of new ones.'

A constructed sense of urgency infuses the industrial agenda with moral purpose, and blocks critique. 'The challenge to feed, clothe and shelter the world's population has never been greater', declared an advertisement published in 1997 by Novartis, a transnational farm chemical company. Industrial strategies marketed by Novartis offered hope: 'New vision and new technology are vital to protect and nourish the crops on which we depend for survival, health and productivity.'[78]

In 1955, the *Agricultural Gazette of New South Wales* praised the work of Australian wheat breeders over the previous half century. The publication implied a relationship between imperial expansion and the rhetoric of hunger, reflecting the global dynamics of power and trade shaping modern agriculture: 'Another James Cook discovering another Australia in that half-century could not have achieved more for a bread-hungry world.'[79]

Australians could help prevent a global food crisis by transferring industrial farming methods to poor countries, agricultural science educator Julian Cribb argued recently. Economic efficiencies achieved by farmers and promises of sustainability offered by the Landcare movement made the Australian system of modern agriculture a valuable export product. 'I can imagine no finer contribution', wrote Cribb, 'which this nation might make to human destiny.'[80]

Talk of ecological disorder in agricultural regions and urgent calls from environmentalists and scientists for fundamental changes to farming systems challenge the moral basis of agriculture. While the dynamics of the global marketplace force farmers to push more and more produce from land, critics are unable to give landholders financial or institutional support to explore alternatives. Conflicting demands foster insecurity, and leave farmers vulnerable. In a recent advertisement, a major agricultural fertiliser manufacturer, Incitec Fertilizers, offered farmers 'Certainty. In an uncertain world.'[81] Clearly, the advertisers engaged by the fertiliser company understood how promises of certainty could tempt farmers.

Australian agriculture is an increasingly risky and uncertain business. The Incitec ad shows a photograph of a farmer in a checked shirt and baseball cap. Under a blue sky, he stands in the middle of a ripe wheat crop, holding several heads of wheat to eye level, inspecting the robust grains. Golden crop sweeps towards the horizon. 'I was here', the advertisement says in bold black letters across the image, beside the farmer and above the bulky sea of wheat. 'I' signifies both the farmer and Incitec Fertilizers. Successful agricultural production, the words and image of the ad suggests, depends on individual will and industrial inputs, not careful and responsive engagement with living systems.

The farmer in the photograph appears content and secure. No other people or species are visible. No disquieting voices rise.

~ ~ ~

One autumn evening I drove to a meeting at Stockingbingal school. Native plant enthusiasts, local schoolteachers and the district Landcare coordinator were discussing ways to protect and tend a remnant of grassy woodland inside the cemetery on the outskirts of the small railway town. A cold front had swept the region that afternoon. Warm air rose from gas heaters towards the high ceiling of the old schoolhouse. Beyond the sash windows and weatherboard walls, gentle rain returned moisture to the earth after the dryness and heat of summer.

Across one of the schoolroom walls, school kids had arranged careful drawings of rare woodland plants encountered inside

Stockinbingal cemetery. Art teacher Sally Last described school excursions to the grassy reserve, where she helped students observe and reproduce the particular forms and colours of woodland grasses and forbs.

Schoolteacher Bill Godman spoke about the involvement of local kids in developing a conservation management plan for the cemetery. Bill had pasted a photograph of flowering golden moth orchids onto a notice to advertise the schoolhouse meeting. He'd encountered the golden moth orchids the previous spring on the grassy few hectares of Stockinbingal cemetery. Bright yellow petals, like the wings of moths in flight, extend outwards and upwards from the green flower stems of the small plants. Botanists have never formally described the particular Stockinbingal variation of the orchid, Bill explained. Similar but different varieties grow in other small town cemeteries beyond the horizon, and on distant roadside reserves.

Before colonisation, the colours and shapes of golden moth orchids and other species shifted by degrees across country. Agricultural development fragmented the intricate patterns of grassy woodland life, silencing the particular expressions of distinctive organisms.

Later that year, with some of my family, I accepted an invitation from a long-time friend to visit her farm near Stockinbingal, to walk through the bush and to see an axe-grinding site in a grazing paddock nearby. Evidence of past Wiradjuri presence across country now fenced and farmed is rarely talked about on the southwest slopes. Landholders tend to keep knowledge of eroding cemeteries, ceremonial stone arrangements, rock art and other surviving traces to themselves and trusted friends.

We arrived at Margot's homestead soon after breakfast. Already the sunshine was hot. Storm clouds built in a bright December sky. Grasshoppers leapt at the windscreen as we drove down dusty tracks through golden paddocks, sheep dogs running behind. As we walked in the forest, ironbark trees and wattles gave only partial shade. Our cattle dog spotted a black wallaby and sped away in futile pursuit. Insects called in the heat.

We returned to the ute and drove further, pulling up beside a fence of rabbit netting and barbed wire on the property boundary. Uphill a short distance, on the other side of the boundary fence, large blocks of sandstone lay scattered along the contour. Margot spent a few minutes looking for the axe-grinding workstation among the summer grasses. She called out and we walked across. On a pale slab of sandstone were eight or nine shallow grooves. Lichen stained each depression, indicating generations of disuse. I noticed ironbark trees, trunks black and fissured, on the stony ridge above us. Downhill, sheep grazed near a dam of muddy water. An elderly yellow box stood alone in the paddock, near a dry watercourse. We walked along the hillside and found other stones bearing grooves and colourful mosaics of lichen.

Climbing back through the boundary fence, we angled our bodies to avoid the rusted wire barbs. Behind us, the sound of an engine carried from the hillside beyond the grinding stones. The manager of the neighbouring farm rode towards us on a motorbike, a sheep dog sitting behind him. Man and dog gazed with suspicion. Margot took the lead, explaining we'd walked up to the ridge to see the trees and shrubs. 'They like the bush', she said. Margot introduced us and we shook hands.

Talk shifted immediately to current agricultural issues: rainfall, another mad cow scare in Japan threatening exports, a possibility of the United States eliminating tariffs on Australian lamb, cattle breeds, embryo transfer technology. No-one mentioned the marked blocks of sandstone resting silently on the grassy slope nearby. The evocative stones, each of us knew, held power to initiate change. Silence defended an established settler stance towards history and place. On the hillside near Stockinbingal, imperatives of export production and agribusiness blocked responses and healing. Powerful traditions suppressed possibilities of dialogue and justice.

revolt

Death and disorder

As shadows lengthened one autumn afternoon, I drove through North Wagga towards the Murrumbidgee River. A dirt road branched away from the bitumen before the bridge, on the uneven floodplain where giant river gums stand. My car dipped into the dusty bed of a dry billabong, Parkan Pregan Lagoon, then up again, onto Pregan Island.

In the 1870s, Mary Gilmore walked to school past here. When she came to write her memoirs, she couldn't remember the Wiradjuri name for Pregan Island, a name her father knew. Before the intensification of agricultural development around Wagga, Wiradjuri people reserved the grassy island as a sanctuary for the guriban, the bush stone-curlew. Sanctuary law, Gilmore explained, allowed pairs of the ground-dwelling bird to nest at the riverside place undisturbed.[1]

Curlews are evocative creatures. Generations ago on the southwest slopes, the gaunt, mottled birds filled the night air with lilting, mournful cries. Wiradjuri stories feature the mirriyuula, a ghost dog with the head of a clever man that comes when curlews wail.[2] The cry of a curlew signals a nearing death, elders told Margaret Tucker, especially if the bird calls from a tree above where people sleep.[3]

Maybe in response to Wiradjuri understandings, one settler on the southwest slopes drew exactly the same meaning from the calls of bush stone-curlews. Banjo Paterson grew up on a property near Binalong, southeast of Harden. He remembered how curlews 'wailed like banshees', female spirits of Gaelic folklore whose cries warn households a resident will soon die. 'My imaginative cousin', wrote Paterson, 'said that they were the souls of lost people asking their way home.'[4]

Many people on the southwest slopes born before World War II hold vivid memories of the bush stone-curlew and its haunting cry. Charlie Stanyer was born in 1909 and grew up on a farm near Illabo. 'They were woeful in the night', Charlie told me. 'They cried.' As a child on a farm at Mimosa, southwest of Temora, George Crooks found a curlew egg and took it to school. 'Mrs McKelvie said it was bad luck', Crooks recalled.[5] Bush stone-curlews were at times thought good omens too. Some people believed the birds wailed and made rare visits when rain was coming.[6]

A half moon began to glow in a hazy sky above Pregan Island and the Murrumbidgee. Pale smoke rose in the distance as farmers took advantage of the calm evening, burning stubble paddocks in preparation for autumn rains and sowing. I walked along the track beneath screeching cockatoos, over broken glass pressed into grey floodplain earth. A riverbank screen of willows, river oaks, golden poplars, red gums and privet blocked views of swirling Murrumbidgee waters. On the eastern side of the island, beyond a sign saying 'Rivturf Instant Lawn', extended a trimmed, green expanse. Two men loaded heavy coils of soil and grass onto a truck.

Distant from the riverbank stood an old river red gum. Final rays of sunlight made leafy branches glow. Sturdy limbs stretched out above floodplain weeds and dry grasses. Now when night falls, the drawn whistles of bush stone-curlews no longer resound against the solid trunk of the elderly red gum.

Curlews were extinct across most of the southwest slopes by

the time I was born. Farming and grazing made life difficult for the ground-dwelling birds. In the western Riverina early in the twentieth century, entomologist Keith McKeown watched curlews trying to divert sheep from nests hidden in the grass.[7] They would run by, trailing wings as if they were broken, or flutter along the ground to suggest a hopeless wound. Sheep were not tempted, of course. Hard hooves smashed countless curlew eggs.

In the 1950s and 1960s, cereal crops and 'improved' pastures replaced wide remnants of native grasses. Farmers applied new technologies and scientific theories to establish high-input, high-output farming systems. Erasure of remaining grassy woodland plant communities doomed the curlews. Farmers removed old paddock trees and fallen limbs to allow the passage of broad machinery. Curlews build nests where woody debris and native grasses offer shelter and concealment. Foxes, dogs, cats and other predators took nesting birds and chicks as logs, decaying branches and perennial grass tussocks vanished.[8]

In the early decades of the twentieth century, Barellan stock and station agent George Gow recorded a dramatic process of local extinctions. Gow had lived in the Barellan district, west of Temora, for thirty years. Before the railway line was extended from Temora in 1908, he managed Moombooldool pastoral station, east of Barellan. Later, as manager of Barellan station, he oversaw the subdivision of the grazing property into farming blocks. Railway transport made shifting grain to urban markets and export terminals easier and cheaper. Agricultural development proceeded apace. Gow observed consequences of ecological fragmentation across the Barellan district as extensive pastoralism gave way to intensive agriculture, the same tumultuous process witnessed by Mary Gilmore and her father around Wagga several decades before.

In the 1920s, Gow published a quarterly booklet to promote his stock and station agency.[9] Articles covered many topics – fat lamb prices, crop competition results, anecdotes of local history.

As settlers established farmland and erased grassy woodland, Gow recorded in his bulletin the decline and local extinction of many plant and animal species. He found the narrowing of biological diversity regrettable. Despite witnessing such destruction, however, Gow maintained faith in the industrial project of agricultural development. Ecological loss and disorder, he implied, were inevitable byproducts of human progress:

> Wheat now takes the place of wool; 100 people can live where one did 50 years ago; matters are completely changed, and it is all for the best.

Before agricultural development, George Gow remembered, late in the nineteenth century, the Barellan district 'teemed' with small animals: bilbies, possums, paddymelons, kangaroo rats and many other species. Some losses of native fauna happened inadvertently. Echidnas were 'rapidly becoming extinct', Gow explained. He once found twelve echidnas in a pit trap designed for rabbits. Emu and malleefowl, plentiful in 'the old sheep days', were now 'driven out by civilisation.' Gow hadn't seen a quoll for decades. Water rats no longer swam in pools along Mirrool Creek.

Deliberate actions brought other deaths. Old papers of the Narrandera Stock Board, Gow discovered, spoke of numerous violent deeds. One season early in the 1890s, a single trapper destroyed almost two thousand wedge-tailed eagles – thought to take lambs – on stony ranges near Binya. As selectors established farms on Cowabbie station near Coolamon, hundreds of brolgas grazed the emerging crops. To meet demands rising in distant cities for grain, farmers spread poison, killing the tall, elegant birds. 'They are seldom seen in our district now', Gow noted.

Diversity in plant life narrowed too. The boree forests – once also a feature of the Stockinbingal and Illabo districts – 'were a pretty sight', Gow remembered. Sheep and cattle find the grey leaves of the graceful, weeping tree especially tasty. Settlers felled the trees to feed stock during droughts. Grubs attacked remaining individuals, 'so that the odd miserable trees left alive to-day do not convey to our minds what a forest of these beautiful trees once looked like.'

Nardoo grew profusely across wet areas in boree forests. Few local young people, Gow suspected, could now identify the perennial fern from which Wiradjuri harvested seed to grind and bake: 'No doubt most of them think wheat was the first thing ever grown in this district from which bread was made.'

To build farmland on the southwest slopes, settlers erased and fragmented grassy woodlands and swamps formerly tended by Wiradjuri. Natural stability and productivity depends on webs of ecological relations, on the interaction of various organisms and elements. Local extinction and decline of individual species made land vulnerable and brought chaotic responses.

In the 1940s, Donald Mackay described escalating patterns of instability and disorder evoked by industrial methods of agriculture in the Wallendbeen district. Born in 1870, Mackay watched the fragmentation of grassy woodland places by agricultural development, a dramatic process that Mary Gilmore, her father and George Gow saw unfold elsewhere in the region. During the early days of pastoralism, Mackay told writer Frank Clune, before the coming of the railway,

> possums and native cats were to be found in nearly every hollow stump. Animal and bird life abounded in the bush; very little country had been cleared, and fallen timber just lay where it fell, giving the animals ample cover. Nowadays, the native fauna have practically disappeared. The country is cleared or ringbarked – thereby making it more productive. On the other hand, the man on the land has to contend against soil erosion, caused by wholesale destruction of timber and careless cultivation. Originally, Australia had no real pests; now we are plagued by rabbits, blow flies, and a multitude of imported weeds. Our natural grasses have been eaten out by overstocking and rabbits, and we have to rejuvenate the soil by planting subterranean clover or top-dressing with phosphate. Fluke in sheep is becoming worse since we shot out nearly all the wild duck, which are Enemy Number One of the fluke-snail.[10]

In the early decades of the twentieth century, Department of Agriculture scientists advised farmers on the western slopes to fallow cropland for an entire year before sowing.[11] Repeated cultivation kept reserved paddocks bare of grasses and weeds. Farmers began ploughing as autumn rains dispelled summer dryness. Later, after steady winter rainfall had soaked cultivated expanses, they hitched tractors or teams of draught horses to sets of harrows. Harrowing groomed paddock surfaces – the metal points of the heavy equipment broke apart clods of earth upended by ploughing.

Farmers harrowed paddocks two or three times during the summer months to maintain a loose 'soil mulch' and to eliminate thirsty weeds. Sometimes, they used disc cultivators to work fallowed land.[12] Pulverised surfaces limited evaporation of moisture absorbed in autumn and winter. Unless drought banished autumn rains the following year, farmers sowed grain into damp and friable paddocks. Unfortunately, the recommended 'long fallow' method made land especially vulnerable to wind and water erosion.[13]

Houlaghans Creek meanders through the Junee district, through paddocks of wheat and sheep, to meet the Murrumbidgee River just below Wagga. Bulky roots of river red gums extend beneath the deep channel of the intermittent waterway. When good autumn and winter rains soak farmland north of the Murrumbidgee, Houlaghans Creek begins to flow southwest.

Max Leitch grew up on a pastoral station downstream from where Houlaghans Creek joins the river. The grazier remembered the devastating effects of a great downpour across part of the Houlaghans Creek catchment in 1914. In a remarkably short space of time, storm clouds had dropped four inches of rain across fallowed wheat paddocks, exposed and vulnerable. Houlaghans Creek churned towards the Murrumbidgee. Swiftly, the river level rose six feet, brimming with 'liquid mud'. Farmland lost vast quantities of red-brown earth:

> The mud killed every living thing in the river. Fish lined the banks

with their heads out of water in the morning, and were all dead floating upside down by lunch time.

Father and the two men employed started pulling some out before breakfast. Some were to be cleaned to eat fresh, some were salted and smoked, and a wagonette load was taken around the farms off the river to be given away. There was one huge Murray Cod that was too big for two men to pull out of the water onto a sand bank. They estimated it to be well over 300 pounds and its mouth was two feet wide. Someone put its head on a stump and it remained there for many years.

The lobsters and shrimps crawled out onto the bank to die in a solid band three to four feet wide and about a foot deep, and also two eels. As children, we had never heard of eels and took Father to see the water snakes in great excitement. Eels are not supposed to inhabit the western watershed. There would have been a ton of dead and dying fish to every one or two hundred yards of river bank.[14]

On parts of the southwest slopes where soils are particularly unstable, trees play a vital role in preventing erosion. 'Saw Ringbarkers doing good work', the manager of Wantabadgery station, southeast of Junee, wrote in his diary in July 1879. 'Engaged Chinamen to burn off dead timber'. Clearing was still happening on Wantabadgery in 1886. 'Saw Peter and showed him how to knock off suckers', the station manager noted in April. And in July: 'White getting on very slowly with scrubbing in Junee paddock.'[15]

Sixty years later, Department of Agriculture inspectors reported the effects of ringbarking on the western side of Wantabadgery station, near Junee:

> Many very bad gullies are evident. This is typical of hilly granite country. The destruction of timber has been excessive, and has been largely responsible for the erosion which is evident.[16]

If remedial action wasn't immediately taken to protect the inland paddocks of New South Wales, warned soil conservation pioneer Sam Clayton in 1931, 'thousands of acres of the best wheat lands in the State will be washed and gullied into barren wastes.'[17] The agricultural scientist found gullies only a few years old on the southwest slopes already metres wide and deep: 'The damage has to be seen to be believed'.

Settlers 'deliberately flouted nature in opening up and developing new country for agriculture', claimed the 1937 edition of *The Land farm & station annual*. Erosion represented 'nature's answer to man's ruthless and reckless challenge', a wounded and rebellious cry.[18]

The following year, the New South Wales government put Clayton in charge of the new Soil Conservation Service.[19] In the postwar decades, Soil Conservation Service workers on bulldozers filled many gullies fragmenting paddocks on the southwest slopes. Contour banks slowed and redirected water flowing downhill.

In the 1970s and 1980s, farmers abandoned the method of repeated cultivation to control weeds. New herbicides and more powerful machines enabled the development of farming systems designed to restore and maintain soil health. Organic content and structure returned to soils, minimising the frequency and severity of erosion events.

Alec Hansen didn't work at the Silver Star café on Thursday afternoons. When a full moon promised visibility for the bike journey home, he rode south from Cootamundra towards a forested range of steep hills. In the 1940s, biologically diverse bushland cloaked the hilly terrain between Cootamundra and Gundagai. There, Alec enjoyed observing and learning about the varied forest plants and animals. He walked through the hills for hours, until he reached a point where the lights of Gundagai became visible.

Later, landholders erased the bushland Alec cherished, back paddocks formerly reserved for occasional grazing. They sowed crops and pastures across bare hillsides, participating in the postwar intensification of farming systems. As years passed, Alec began to notice infestations of pest insects on isolated paddock trees throughout the Cootamundra district. Before the clearing of forested country south of Cootamundra, he told me, predatory insects flew or were blown by winds from the hills into paddocks where they restricted leaf-eating insect populations. Now ecologies were disordered and unstable, natural patterns fragmented. Paddock trees sickened and died.

Land appeared to steadily turn dry as rural colonisation progressed. Springs reliable in the early 1870s no longer flowed in ranges west of Barellan, near Binya, George Gow noted in his quarterly bulletin. The springs, a pastoralist told him, had 'silted up' and disappeared.[20]

Other chroniclers of ecological change on the southwest slopes recorded a growing scarcity of moisture. In the Young district, east of Barellan, James White, Sarah Musgrave's uncle, established sheep runs where he found continuously flowing springs. In time, Musgrave explained, stock trampled and compacted the ground so much that water stopped rising.[21]

Back in the 1860s, remembered a Junee district resident, the dense and widespread cover of native perennial grasses ensured that water rarely flowed down hillsides.[22] As landscape ecologist Christine Jones explains, most native grasses form vase-shaped tussocks to capture and draw rainfall underground.[23] Friable and rich in organic matter, soils across the western slopes absorbed and held great quantities of moisture.

Many processes kept soil loose and absorbent. Wiradjuri unearthed yams and other edible tubers and roots with digging sticks. Diverse and abundant populations of small ground-foraging mammals probed and scooped the grassy woodland surface in search of insects, roots and fungi. Nocturnally active, the creatures buried fallen leaves, bark and other materials as they churned spaces between grass tussocks, bolstering the organic content and moisture-holding capacity of soils.

Patterns of water flow changed as perennial grasses, small mammals and Wiradjuri people disappeared. No longer did the slopes act like great sponges, holding moisture and feeding springs and swampland. Rainfall washed away down bare, compacted hillsides. Gullies scarred the land.

Moisture vanished from the Junee district as colonists erased grassy woodland adapted to capture and hold water. 'When this country was all bush we used to have fogs and mist all the winter', recalled farmer Charlie Stanyer.[24] Early in the twentieth century, his parents cleared paddocks near Illabo, northeast of Junee, and established a farm. 'It was then just virgin forest', Charlie explained. By the outbreak of World War II, the Stanyer family farmed wheat across more than six hundred hectares.

'The whole country looked to me beautiful', wrote George Seymour of his first trip through the Cootamundra and Junee districts in the 1860s, 'it seemed to me that no land was so fair.' Fences were almost absent, 'and with kangaroo grass covering the hills, and the river hemmed in with rushes, I pictured it the most favoured spot on earth.' Only bullock teams carried supplies and produce, Seymour explained, because the land was 'too boggy' for horse transport.

As a worker on Wantabadgery station near Junee, Seymour observed processes of ecological change imposed by pastoralism. 'The kangaroo grass was not eaten out by the rabbits as some people think', he wrote, 'the sheep killed it.'[25]

When pastoralists Thomas Hammond and Richard Gwynne took over Junee station in the 1850s, most of the area was 'open forest country', Hammond's son explained,

> and coated chiefly with kangaroo grass. Later on when the sheep rapidly destroyed the kangaroo grass, the country went through a period equivalent to a drought, although the rainfall was good, because other varieties of grass had not arrived to take the place of the kangaroo grass.[26]

'Now, Dame Nature is peculiar', noted George Gow. 'If you interfere with her, she is like a boomerang; you never know how she will rebound and hit you, and to what extent.'[27]

In January 1905, when my great aunt was only a few years old, a frightening bushfire swept the Riverina. As an elderly woman, she remembered taking refuge from the flames inside a cellar beneath the family homestead near Tarcutta Creek. Perhaps it was her earliest memory. The fire began in the Jerilderie district, almost two hundred kilometres west of Tarcutta. By nightfall, the foothills of the Snowy Mountains were burning. People and stock perished as dry, gusty winds fanned the conflagration. Flames consumed wheat crops, livestock, homesteads and people. At one stage, the fire front stretched eighty kilometres from the Murrumbidgee River near Tarcutta southeast to Tumbarumba. Twenty-seven thousand sheep grazed Tarcutta station. Half survived.[28] Fire destroyed the bridge over Tarcutta Creek, and the woolshed on Tarcutta station burned to the ground.[29]

Ecological changes brought by colonisation promoted intense and destructive bushfires. Removal and decline of native vegetation left country exposed to desiccating summer winds.[30] Trees, shrubs and grasses turned dry and flammable. As hillsides lost capacity to capture and hold water, plants could no longer draw upon reserves of soil moisture. Dry leaves and branches fed hungry flames. The widespread extinction of small mammal species worsened the situation. A brush-tailed bettong performs up to a hundred diggings every night. One bettong can churn six tons of earth annually. As they forage in the ground, bettongs and other creatures bury great quantities of fallen bark, leaves and other combustible materials.[31]

On the southwest slopes and plains, wildfire interacted with a range of other factors to generate unexpected, extraordinary responses. When fire swept ringbarked land, thick stands of cypress pine, eucalypt and other woody plants rose to confound settlers. John Holloway, owner of Moombooldool station near Barellan, once contracted a team of Chinese ringbarkers and scrubbers to clear sixty thousand acres of box trees.[32]

On Moombooldool and throughout the Riverina, grasses flourished as ringbarked forests died. When summer wildfires ravaged cleared paddocks, and winter rains soaked blackened earth, millions of woody plants sprouted. Few small native mammals remained to graze and eliminate emerging seedlings. 'Thus in a few years', explained Gow,

> the last state was worse than the first; for, instead of open forests through which you could in many places ride at a gallop, dense scrubs, mostly of pine, appeared; and as it grew it was so dense as to become difficult to ride through.[33]

The Fisher family took up a farming block in 1882 on Yarranjerry station, west of Temora. Elizabeth Fisher tied bells to her children so she could find them if they vanished into the enveloping scrub.[34]

Another early settler in the same district remembered the sounds of wild cattle stampeding. Beasts snapped trunks and limbs of young trees as they crashed through the rising bushland.[35]

Scientific and technological efforts to harness rural places for export-oriented production brought unexpected and unfortunate consequences. In time, strategies to control complex and dynamic natural systems often proved misguided. As sociologist Nigel Clark explains,

> the Enlightenment project – the quest to impose order and intelligibility on the world – has ultimately exacerbated the very uncertainty it sought to abolish.[36]

On the southwest slopes, industrial power fragmented ecological connectivity and undermined the resilience, stability and natural productivity of land. Today, global economic structures and government policies promoting competition maintain destructive styles of engaging with rural places. Narrow, productionist goals fail to incorporate the complex, enduring needs of local ecologies and human communities. Subsoils turn acid, salt scalds widen, paddock trees die, people leave. Farmland is pushed, and natural patterns continue to unravel.

~ ~ ~

Currawong is a farming and grazing district north of Harden, southeast of Young. James Roberts established Currawong and Milong squatting runs late in the 1820s. Not twenty years old, Roberts 'must have been a remarkable man to carve out and hold two runs', thought local historian Richard Littlejohn.[37] Convict servants and free workers helped the young squatter tend stock, grow hay and cut eucalypt slabs to build huts.

Decades later, selectors arrived with steel ploughs and draught horses. One winter morning I met Currawong district farmer Ted Brown at his homestead. Ted's family selected land here in the 1860s, when his grandfather was a boy. The Brown family first lived in tents erected on the undulating 130 hectares they called 'Rosevale'.

Ted's property is much bigger than the block selected by his forebears. In the global marketplace, Ted explained, local farmers today need at least a thousand acres (over four hundred hectares) to make a profit.

Ted described a flat area thick with spike rush, a plant indicating wet and often salty soils. When his grandmother's family arrived at Currawong, they built a home where the rush now grows. The look of the place shows how much the country has changed, he told me. Stories handed down in the family record a dramatic process of loss. Ted's grandfather remembered koalas grunting in trees at night. Other animals are no longer heard or seen around Currawong.

Years ago, Ted learned from an elderly man, many kangaroo rats lived in a particular paddock. The small mammals built mounds of sticks and grass, then burrowed inside to nest. Since his childhood in the 1930s, Ted has noticed the disappearance of goannas, possums, babblers and banded lapwings. Large flocks of currawongs no longer descend from the mountains in winter.

Some species seem to like the changes modern farming has brought to local ecologies. Once rare in the district, grey kangaroos, galahs and crested pigeons are now common. Ted sees growing numbers of superb fairy-wrens in paddocks and inside the homestead garden. I mentioned bush stone-curlews. Ted hasn't seen or heard one for at least forty years. He remembers the mottled birds crouched under trees and beside fallen timber as he rode to school.

I asked Ted to describe the cry of a curlew. As a child, he often mimicked the eerie sound to induce the birds to call. Inside the homestead where we sat talking by the fire, Ted whistled a rising, mournful wail.

Few of the hillsides around Currawong remained cloaked in grassy woodland by the time Ted was born. Ringbarked trees stood in some paddocks. Sheep grazed native grasses among fallen limbs and grey trunks.

Farmers bulldozed most local stands of dead trees soon after World War II. In those days, Ted explained, the heaviest crops grew on 'new ground', land cleared and ploughed for the first time. Today, inorganic fertilisers and new plant varieties allow the harvesting of even bulkier yields from paddocks cropped repeatedly. Ted has planted thousands of trees across Rosevale since the 1950s. 'They make good shelter for the stock and they make the country look a lot better', he told me. River oak and river red gum saplings thrive beside gullies and across places too salty and wet for cropping.

Local farmers are sowing deep-rooted perennial pastures, Ted explained during a tour of Rosevale, species able to soak up more rainfall than shallow-rooted annual crops and pastures. We discussed other strategies designed to make farming more sustainable. Ted mentioned the widespread application of lime to reverse soil acidification. But future problems, he feels, will inevitably arise. Despite generations of family engagement with the natural patterns of Currawong, Ted faces the land with uncertainty. 'You do one thing', he said, 'and another thing comes up.' Widespread erasure of grassy woodland life, it seems, have blocked opportunities to sense and understand ecological dynamics: 'We're only just beginning to know the land, I think.'

Ted stopped the farm ute inside a woodland remnant. He built a fence around the shady place twenty years ago. By accident, eight hectares of yellow box and red gum trees, kurrajongs, chocolate lilies and red grass had escaped generations of clearing and cultivation. The red gums looked sickly when Ted and I walked under them. Dense populations of psyllids and other leaf-eating insects cloaked branches. Greening Australia had recently helped Ted sow seeds of drooping she-oak, several varieties of wattle, and other shrubs. Beside our feet, minute black and brown seeds sprouted amid grass tussocks and vigorous weeds. As young trees and shrubs grew into rare habitat, Ted explained, small birds would return to feast on pest insects now attacking the red gum leaves above us.

Beside yellow box trees and regenerating wattles inside the enclosure, a dam wall holds water where two slopes meet. Ted pointed to a raft his grandchildren had made from surplus timber and plastic drums. 'They swim and camp here', he said. The remnant of grassy woodland contrasts with surrounding cultivation and grazing paddocks. Inside the reserve, diverse life offers a sense of stability and comfort. With our backs to the industrialised, open farmland around us, we gazed into trees and examined the ground for seedlings. Sometimes, Ted drives to the shaded, lively place to light a fire and cook lunch. 'It's very peaceful here', he told me, 'out of the wind.'

Survival and revolt

A worn track leads from the bitumen road towards sheds of corrugated iron and white cypress pine. Beside flat paddocks and Pinchgut Creek, west of Cootamundra, the old Retreat woolshed looks monumental. Several yellow box trees shade the bare, compacted space between the sheds and the cattle yards, sturdy limbs holding leafy crescents to the sky. Bedrooms in the derelict shearers' quarters are stacked with chemical drums and bags of fertiliser. On hot December days, the elderly yellow box trees flicker shade across grain-laden trucks leaving paddocks for railway silos.

Recently, I noticed a dump of agricultural lime – brilliant white in midday autumn sunshine – between the stately box trees near the iron shearers' quarters. Contractors later dusted Retreat hillsides and creek flats with the pulverised limestone. The highly alkaline and especially reactive mineral is applied regularly to farmland to counter soil acidification. Australian farms are coated with two million tons of agricultural lime each year. According to Environment Australia, almost seventy million tons may be needed to reverse acidification and keep farmland productive.[38]

Concern about soil acidification deepened in the 1980s. As most agricultural products are slightly alkaline, the consistent export of primary produce from paddocks steadily increases soil acidity levels. In the later decades of the twentieth century, new high-input, high-output farming systems accelerated the acidification process. Artificial fertilisers and leguminous plants like clover and lucerne substantially bolster soil nitrogen levels. Excess nitrates filter down soil profiles, causing increased acidity.

Soil acidification limits ecological wellbeing and productivity in many ways. Earthworms and beneficial micro-organisms can't live in highly acidic soil. Calcium, magnesium and other elements necessary for plant health become less available as acidification intensifies, while aluminium, manganese and other toxins become more accessible to plants. Soil structures deteriorate and plants lose vigour, making farmland more vulnerable to erosion. Water escapes below the sickly roots of crop and pasture species. By restricting the growth and water consumption rates of plants, soil acidity promotes dryland salinisation. Rising watertables bring dissolved salts to the surface, compounding damage to land.

In the 1980s, sheep and cattle enterprises gave way to broadacre cropping across many properties on the western slopes. Shifting dynamics of domestic and global economies made intensive farming more profitable than wool and meat enterprises.

Late in the nineteenth century, Retreat was a larger holding, a station devoted to the grazing of merino sheep for wool production. Edward Donnelly ran the property on Pinchgut Creek in conjunction with Borambola, his home station beside the Murrumbidgee River near Wagga. Donnelly also held grazing stations on the southern tablelands, beside the Darling River in western New South Wales, and in the Snowy Mountains.

Responsible for a pastoral empire of almost a million acres, Donnelly suffered badly during the major economic recession of the 1890s. As his debts soared, creditors Goldsbrough Mort and Company stepped in to oversee the management of his stations.[39]

The firm made regular inspections. 'A good deal of the country has not been very long ring-barked, & the trees are still throwing out suckers, which require constant attention', property inspector John Ross observed at Retreat in 1894. 'As regards suckers and seedlings', he reported two years later, 'these have been well looked after, and there is hardly one to be seen on the place.'[40]

Another problem emerged in one ringbarked paddock. Sifton bush was 'making great headway', and required control. To suppress the plant, the resident manager of Retreat planned to cut tracks through the expanse of sifton bush in summer and set fire to the paddock.

Sifton bush, *Cassinia arcuata*, favours disturbed ground. The small plant with aromatic leaves and brown flowers is common in quarries and mining areas. In his book *Weeds in Australia*, published in 1976, agricultural scientist Charles Lamp noted its 'aggressive properties'. NSW Agriculture lists sifton bush as a noxious weed in the Harden and Young districts, legally obligating farmers there to eliminate the plant.

Alec Hansen pointed out the shrub when we met at Jindalee State Forest on a warm spring morning. For decades, Alec has observed with deep interest the intricate and shifting patterns of nature on the southwest slopes. He doesn't consider sifton bush a weed. The hardy, drought tolerant shrub colonises cleared areas. Fallen leaves cloak soil. Roots hold earth. Thickets of sifton bush offer shade and cooler temperatures, allowing soil microorganisms to reproduce and thrive. As ecological stability returns, Alec explained to me, dense communities of sifton bush thin and disappear. Eucalypts and other species grow. Sifton bush heals disturbed, wounded places.

A deep-rooted perennial, sifton bush helps slopes resist the erosive power of wind and water. In 1902, botanist Richard Cambage found sifton bush 'very plentiful' in the Temora district.[41] He saw the shrub spreading across bare wheat paddocks.

Sifton bush and wheat like the same growing conditions. Wheat and other annual crops are direct descendants of plants adapted to swiftly colonise areas devastated by natural phenomena like floods and wildfire. Such plants need big, easily detached and sturdy seeds to survive the next catastrophe.[42] To provide suitable conditions for shallow-rooted, annual crops, farmers must apply industrial devices each year to replicate catastrophe.

When persistently disturbed, land is vulnerable. Despite the ability of late twentieth century cropping systems to bolster soil structure and organic content, denuded and ploughed paddocks remain exposed to extreme weather events. In March 2003, storms drenched Cootamundra farmland and broke an extended dry spell. Soil washed from bare hillsides. Cootamundra Shire Council sent a grader west along Dirnaseer Road, towards Retreat, to scrape damp topsoil from bitumen surfaces.[43]

The healing spread of sifton bush is one example of local ecologies asserting agency in the face of industrial methods of agricultural production. Unfortunately, few settlers heeded the messages offered by the vigorous plant. As the erosion events of March 2003 in the Cootamundra district show, land becomes vulnerable when ecological complexity and connectivity is fragmented and lost.

Other natural agencies have expressed even louder challenges to the industrial mindset of modern agriculture. 'ONE INSECTICIDE PROTECTS A WHOLE FARM!' declared a 1963 newspaper advertisement for the chemical product Malathion.[44] Until the early 1970s, farmers and grain handling authorities relied on the organic phosphorus insecticide to protect animals and stored grain from insect infestations. Widespread reliance on the chemical product gradually bred resistance into insect populations. In the summer of 1973, unwelcome rains fell across the western slopes as farmers tried to harvest crops. The Grain Elevators Board of New South Wales agreed to accept relatively moist grain, despite a greater risk of insect infestation. Board staff struggled to contain insect numbers inside storage facilities. Malathion, they discovered, no longer worked.[45]

The following year, the Grain Elevators Board chose a different grain insect poison, bioresmethrin. Chairman Peter Druce admitted that bioresmethrin offered only 'breathing space'. In the longer term, the board needed to find means other than poisoning to protect stored grain. The dynamic ability of insect populations to resist and subvert human efforts at mastery, Druce realised, ensured the eventual failure of any chemical control strategies.[46]

Mounting problems with insect resistance and the realisations of Druce and others seemed not to undermine general faith in the ability of humans to master living systems for primary production. Many people still thought nature essentially knowable and controllable, a belief on which powerful industrial interests depended.

In the final decades of the twentieth century, agricultural scientists and chemical corporations initiated what political scientist James Scott calls a 'biological arms race' with dynamic living systems.[47] 'Will smart insects inherit the earth?' asked the Wellcome chemical company in 1985.[48]

> Insects adapt quickly, and are clever and hard to kill. Look how the Lesser Grain Borer is breeding strains that are resistant to

organophosphorus treatments, and some Flour Beetles and Saw-tooth Grain Beetles are adapting to maldison and fenitrothion. The Wellcome Solution is smarter: BRM together with internationally proven Reldan, will control all the pests that will have a go at your stored grain.

As regular insecticide applications select for resistant insect strains, reliance on herbicides to destroy weeds eventually breeds troublesome populations of resistant plants. At the end of the twentieth century, Australian farmers found once powerful herbicides increasingly ineffective. Silver grass, rye grass, barley grass, wild oats, wild radish, capeweed, Paterson's curse and other weeds developed immunity to herbicides.

The list of herbicide resistant weeds grows each year. Resistance threatens the productivity of modern farming systems. Heavy infestations of wild oats can slash cereal crop yields by seventy per cent.[49]

Herbicide resistance of annual ryegrass is especially problematic. Ryegrass was once a popular pasture. As grazing gave way to cropping on many farms later in the twentieth century, annual ryegrass became unwanted, a weed competing with crops for moisture and sunlight.

Annual ryegrass spreads profusely and holds remarkable genetic variability.[50] Populations of ryegrass are likely to contain some plants with resistance to different herbicides. Repeated spraying fosters the rapid spread of herbicide resistant annual ryegrass and other resistant plants. As herbicides kill neighbouring plants lacking resistance, resistant individuals thrive, set seed and spread.

Australian settlers understand the power of nature to resist colonial demands. One definition of 'bush' in *The Australian national dictionary* is 'country which has not been settled or which has resisted settlement.' In 1947, philosopher Max Horkheimer described rebellious responses of natural systems to industrial strategies of control as 'the revolt of nature'.[51] Soil acidification, dryland salinisation, the spread of sifton bush, plant and insect resistance to chemicals, and other phenomena active across Australian agricultural regions may be considered expressions of revolt by natural systems against the modern quest for mastery over farmland.

Mindsets of industry and agricultural science promote particular changes to nature and human society. 'However', as Vandana Shiva explains,

> since both nature and society have their own organisation, the superimposition of a new order does not necessarily take place perfectly and smoothly. There is often resistance from people and nature, a resistance which is externalised as 'unanticipated side effects'.[52]

Natural forces and living systems are complex and dynamic, not simply passive and controllable. As the demands of industrial, export-oriented agriculture intensified in the second half of the twentieth century, natural agencies responded vigorously to scientific and technological methods of farmland management. Strategies to eliminate ecological barriers to the maximisation of primary production evoked bigger challenges, louder expressions of revolt. New, powerful technologies and more strident interventions escalated the scale of natural responses.

Revolt by the living systems of agricultural regions was inevitable. As environmental historian Carolyn Merchant observes, 'the tighter the rein, the greater the potential for rebellion.'[53] Today, agricultural and environmental scientists help farmers find 'solutions' to ecological disorders, to phenomena seen as further barriers to production. Perhaps a deeper process of reflection and reform is needed to bring enduring regeneration to rural places. In modern agriculture, dominant notions of 'natural resources' and 'management' suggest the persistence of problematic industrial mindsets.

Drawing on the work of Horkheimer, Merchant identifies two dimensions of the revolt against the industrial domination of 'external nature' – land and other species – by 'internal nature' –

our human imagination and will. One involves the ecological rebellion of external nature. Agricultural examples of revolt by the living systems of external nature include soil acidification and chemical resistance by plants and insects. The second dimension of natural rebellion Merchant identifies is revolt by humans themselves – by our internal natures – against modern efforts at mastery. Industrial and economic 'rationality' demands repression of human emotion and sensuality. To achieve mastery over land and other species, people must adopt narrow, constrained understandings of the human condition.

On the southwest slopes, some colonists rebelled against the self-limiting demands of industrial culture. Late in the nineteenth century, ecological wounding imposed by agricultural development disturbed Mary Gilmore. The poet recorded her personal distress and the pain of other species in 'I Saw the Beauty Go'.

> I saw the beauty go,
> The beauty that, in a stream,
> Flowed through the breadth of the land
> Like the fenceless foot of a dream.[54]

Mopokes called on moonlit nights as farmland spread, 'and the curlew made her pleas.' Seeking mastery, agricultural colonists refused to hear or respond to the voices of the land. Settlers erased components of rural places considered useless or troublesome:

> I saw the beauty go,
> The beauty that could not be tamed;
> But before it went it looked at me
> With the eyes of the maimed.

Colonisation demanded avoidance of intimacy with the native life of the land, with the web of lively relationships erased and disordered by farming practices. Meeting 'the eyes of the maimed' threatened to undermine the political and industrial project of agricultural development. Imperial and economic imperatives disabled negotiation and compromise between settlers and places. Ringbarked forests died. Steel ploughs buried grassy swathes.

Cultural historian Paul Carter describes the process of 'ungrounding' necessary for colonisation to proceed.[55] 'To be groundless', Carter explains,

> was to masquerade as being everywhere: the colonial mind was a citizen of the entire intellectual world, atopic, occupying the transcendental plain of its own reason.

Colonists imagined themselves as physically and emotionally divorced from local particularities of nature. 'Essentially, to be ungrounded was to lose touch with one's human and physical surroundings; it was to become an echoing shell, an antenna eye.'

Carter considers it unsurprising that these 'unnatural circumstances' induced nostalgia and longing for the colonised and distanced 'Other'. On the southwest slopes, colonists appeared to yearn for cultural and emotional connectivity with the same grassy woodland species and natural patterns they devalued and destroyed. Perhaps, as Carter suggests, few settlers were fully committed to the powerful agendas of colonisation. In the 1930s at Junee, the Jeffs Brothers flour milling company named a line of self-raising flour 'Curlew' after the bush stone-curlew.[56] Across the region, the ground-dwelling bird disappeared as wheat paddocks replaced its grassy woodland habitat. Flour bags showed a lone curlew beside tussocks of grass.[57] In 1937, a plant breeder at the Temora Experiment Farm likewise named a new strain of wheat 'Curlew'.[58]

After World War II, as agricultural systems intensified, wheat breeders in New South Wales named new wheat varieties after other birds threatened by broadacre farmland development: Rosella, Plover, Thornbill, Brolga, Currawong, Pardalote, Babbler and Diamondbird.[59] In a similar fashion, oat varieties grown today in southern Australia are named after native animals dependent on fallen timber and grassy woodland plants for survival: Echidna, Bandicoot, Wallaroo, Quoll, Glider, Possum, Bettong and Potoroo.[60]

Perhaps the naming process indicates subliminal recognition of inescapable ties between human bodies and enveloping natural domains. Did western assumptions of disembeddedness and industrial quests for transcendence evoke haunting desires to acknowledge human bonds and obligations to wider living systems?

An illuminated manuscript prepared in 1925 by artist Gordon Nicol presents Banjo Paterson's poem 'Song of the Wheat' amid bright and colourful designs.[61] The poem celebrates the spread of agricultural settlement across the grazing estates of the southwest slopes. 'Yarran and Myall and Box and Pine – 'Twas axe and fire for all', wrote Paterson. On the manuscript, Nicol painted a kookaburra, waratah blossoms, different varieties of wattle, correa, violets, grain-filled heads of wheat, a bunch of grapes, and other native and imported specimens.

As well as implying a longing for life destroyed, the kookaburra, wattle and waratah were also popular floral motifs of New South Wales and the Australian nation. During the 1920s and 1930s, farmers on the western slopes grew Waratah, Federation and Free Gallipoli wheats. The names of these and other wheat varieties reflect the state and national imperatives driving agricultural development and ecological change, cultural and political forces against which land and people have shown varying degrees of rebellion.

On the southwest slopes, revolt by settlers against the demands of industrial and national development sometimes took more explicit forms. Catholic priest Joseph Dwyer moved to Temora in 1912. Travelling across the Temora parish to perform Mass in small churches and homesteads, the amateur botanist marvelled over plants indigenous to the region. He led services inside Redesdale, the Cooney family homestead northeast of Temora on Bland Creek. Mary Morton (formerly Mary Cooney) remembered Dwyer as 'a very outgoing sort of a man', she told me. In Redesdale paddocks the priest picked billy buttons, chocolate lilies and bluebells to add to his native plant specimen books. 'He loved the beauty of Nature's flora in Australia', Dwyer wrote of himself, and thought 'nowhere in the world has the God of Nature spread such loveliness to delight the eyes of mortal man than here.'[62]

The priest took Catholic school classes on botanical excursions into bushland near Temora. Dwyer noticed a distinctly different red gum growing on stony ridges. Unlike other red gum species indigenous to the southwest slopes, the trees Dwyer found grew in mallee form, and had relatively narrow leaves.[63] Joseph Maiden and William Blakely of the Botanic Gardens in Sydney named the red gum species *Eucalyptus dwyeri*, commonly known today as Dwyer's red gum or Dwyer's mallee gum.[64]

In 1921, when he was Bishop of Wagga Wagga, Dwyer published a local floral survey in the *Australian Naturalist*, a popular botanical journal.[65] Across a wide area on either side of the railway line between Temora and Wyalong, he recorded 337 different plant species. As Temora historian Laurel Thompson observes, an extremely narrow range of crop and pasture varieties now cover the land Dwyer surveyed.[66] His floral survey is a rare and valuable record of richly diverse plant communities erased by agricultural development.

Broadscale clearing troubled Dwyer. He feared declining rainfall and shortages of good milling timber.[67] Others in the Temora district shared his concerns. In 1931, local historian Watson Steele noted the loss of 'beautiful cypress-pine belts' once stretching between Barmedman and Coolamon, west of Temora. Such extensive destruction of pine forest for agriculture was misguided, Steele argued, 'for not only has it become very valuable, but stock require shade and shelter, and it retains a certain amount of moisture in the soil.'[68]

Joseph Dwyer wished for settlers to embrace native plants:

> It is a pity the gardens and parks are not planted with many of the shrubby wattles, which adorn our wild bushlands, but will soon be as extinct as the dodo.

There were many 'beautiful shrubs' in the region suitable for garden plantings, like *Acacia cardiophylla*, the feathery-leaved Wyalong wattle, Dwyer noted in 1933, 'but lovers of the wild things should try to propagate some of them before the vandals burn them all.'[69]

~ ~ ~

Bob Glanville lent me a tape recording of his great uncle, Arthur McGuiness, talking in the 1960s to ethnologist and linguist Janet Mathews. I played the tape one winter morning while working in the garden. Mathews, it seemed as the tape began to play, hoped to secure on tape fragments of language and culture before the old Gudhamangdhuray clansman died.

On the recording, Arthur McGuiness said he was born and raised at Cootamundra. The place is his 'native home', his 'ngurambang'. In Wiradjuri, 'ngurang' means 'camp' or 'home'. 'Ngurambang' is formed when the suffix '-bang' is added to a modified form of 'ngurang'. The suffix serves to intensify the meaning of the word it joins. Ngurambang is spiritually deep terrain, belonging place, country. When people are conceived they emerge from inside their ngurambang, out into the visible world. People return inside their ngurambang when they die.[70]

Similar words with similar meanings exist elsewhere in Australia, reflecting bonds of shared culture across the continent. Pintupi of the western desert, for example, say 'ngurra' for 'camp', 'country', or 'place'.[71]

Late in the interview, Mathews asks McGuiness to sing a song. He is reluctant to perform alone, without accompanying singers and instruments. But Mathews is persistent. McGuiness begins singing a song his father taught him as a boy.

On a ladder pruning a grapevine, I stopped work and gazed over red gums and fence lines, across the upper reaches of Muttama Creek. The deep and resonant voice of Arthur McGuiness rose steadily from the tape recorder, a shifting rhythm. In the valley below, winter sunlight dissolved morning fog. The old man sang only a few lines of a song perhaps composed in firelight beside Muttama Creek swampland. The song is about a wicked and treacherous boy, McGuiness tells Mathews. 'I would love to,' he begins, then stops. 'My sister, she's dead and gone. She could sing.' Mathews encourages him to sing another song. McGuiness resists: 'No, no really I couldn't.' The recording ended abruptly but the tape played on, clicking and hissing into cool air.

One warm spring afternoon several years later, I parked beside the railway line near the eastern edge of Cootamundra. River red gum saplings grow there between the road and the railway, where rainfall pools. I walked under the railway bridge, sturdy enough to support trains heavy with grain and other freight. Decorated in graffiti, thick slabs of concrete made the shaded space cooler. My dog splashed into Muttama Creek. Algae and plastic rubbish floated in murky water. A short distance upstream, I passed barbeques and picnic tables in a public park beside a waterhole. Nearby, children played on a tractor and hay baler, antique farm equipment painted green, yellow and red. The tractor, I noticed, is a Massey-Harris, manufactured in Britain or Canada and exchanged for primary produce of inland Australia. Together, the railway bridge, polluted creek and old equipment spoke of a global dynamic of history and ecological change.

Behind the former district hospital, under the shade of weeping willow, orange nylon rope tied to the wire netting of a submerged yabby trap lay alongside water couch. Elm, prunus, willow and phalaris formed a screen on the opposite bank. Further along Muttama Creek, near the public swimming pool, I noticed earth cracked and dry, seemingly a remnant of swamp bed. Earthworks and stormwater pipes ensure winter rains rarely saturate parts of Cootamundra where swampland once extended, the lost watery expanse Gudhamangdhuray clanspeople reserved for turtles. Further upstream I walked away from Muttama Creek. Near the high school, a footbridge crosses a wide floodwater drain bulldozed across the flat. When heavy rain falls, the grassy depression leads excess water straight towards the creek, away from homes, gardens and public buildings.

My dog and I jumped a low fence of wire netting and set out across the high school oval. A band was rehearsing in an upstairs schoolroom. The sound of drums and electric guitar reverberated across the dry sports field. I sat on a wooden bench and my dog panted in the shade of a red brick wall. The strident voice of a young woman, the lead singer, carried throughout the spaces of the school. She sounded angry and restless:

> Something has to change, it's not a burden any one can bear. It's not enough! I need more! Nothing seems to satisfy! I don't need it. I don't want it. To breathe, to feel, to know I'm alive!

The song performed by the high school band seemed linked, somehow, to the erasure of the turtle swamp, and to the recorded song performed long ago by the elderly Gudhamangdhuray clansman Arthur McGuiness. The voice of the old man, I remembered, seemed to rise from the valley, from Muttama Creek swampland. The voice of the young woman seemed to come from somewhere else, from an unnatural space between settler culture and local earth.

Industrial technologies – railway lines, bitumen roads, Murrumbidgee water pipes, fibre optic cables, steel fences – are imposed over the terrain of the southwest slopes. Connectivity between people and the dynamic life of places is denied, responsive and careful relations blocked. Few stories or songs acknowledge the distinct patterns of rural places and the reality of human embeddedness in living systems. Beside Muttama Creek, the streetscape of Cootamundra appears to hover above the land. From the space in between, sometimes, there comes a yearning, natural cry of human revolt.

regeneration

Healing

Across the particularly fertile and productive region of the southwest slopes, few places are both relatively unmarked by industrial agriculture and freely open to the public. In 1893, the Department of Lands set aside a steep, forested area beside Cootamundry Creek as a recreation reserve for the people of Cootamundra. Now known as Pioneer Park, the granite porphyry hills of the reserve are cloaked in biologically diverse bushland. For townspeople without access to rural properties, the forested expanses of Pioneer Park are especially valuable. Alec Hansen first walked across the stony hills in 1946 while visiting family in Cootamundra. The following year he found a job in the district, and moved from his childhood home on the southern outskirts of Sydney.

Alec and I walked through Pioneer Park one spring morning, past stringybarks and spearwood wattles. We stopped whenever Alec noticed a particularly interesting plant or animal. As we walked uphill, Alec spotted a common fringe myrtle, *Calytrix tetragona*, and we left the track to take a closer look. When the shrub finishes flowering, Alex explained, *Calytrix* petals fall away, leaving an impressive display of red calyxes on the shrub. Eventually the calyxes descend too, each with a seed attached. As rain falls and humidity rises, threads attached to the calyxes absorb moisture, causing them to twist. The turning action pushes *Calytrix* seeds into damp soil. Germination begins.

regeneration

We kept walking in the sparse shade of red gums. During World War II, the Australian Air Force made a base here, Alec told me. Workers cleared trees and shrubs, causing blankets of loose soil to wash downhill. When the Air Force abandoned the site after the war, the shire council leased Pioneer Park to local graziers. Some years later, when the council decided to remove livestock, bushland regenerated swiftly. Trees cut down a generation before grew back from shoots emerging at the bases of stumps. Seeds of forbs, grasses, shrubs and trees high on the granite hills washed and blew onto bare spaces below. Over decades, a diverse and beautiful forest community returned.

We stopped. Alec spoke gently to a swamp wallaby paused in the shade nearby, watching us pass. He whistled softly to reassure the creature. We recommenced walking across the hillside through a stand of black cypress pine. Cootamundra Shire Council recently built fences inside Pioneer Park. According to the council, Alec said, the fences were to exclude young people on trail bikes. Cattle now grazed the fenced areas inside the reserve, supposedly to reduce fuel loads and the risk of bushfires.

Alec saw much damage caused by the livestock. Cattle nibbled away and pushed over shrubs. He suspected cattle could smell the tubers of greenhoods and other underground orchids. The beasts unearthed and ate many of the fleshy orchid tubers. The scent of cow dung now lingered where crisp bush smells once prevailed.

Alec doesn't see forested areas like Pioneer Park as a sort of 'wilderness' from which human activity should be banned. He pointed out different paths taken by a horse-riding group through the park. Alec was sorry town kids could no longer ride trail bikes in the bushland reserve. No harm was done if the young riders stick to tracks protected from erosion by earthworks. 'And you get some bush devotees by allowing them in', he told me.

Moving downhill, Alec sighted a wood swallow and stopped. Great flocks of the smoky brown birds, he explained, journey annually from northern Australia to nest in southern forests. We reached a basin at the foot of the slope and entered an established grove of Cootamundra wattles. Years ago, Alec came to this sheltered place with a nephew who was leaving town. They planted a wattle together. Alec repeated the practice when other family members left the district. The small trees are now mature and spreading. When Alec visits the acacia grove in Pioneer Park, each wattle reminds him of a relative. As we stood amid the Cootamundra wattles, Alec again abruptly stopped talking. A brown treecreeper mounting the trunk of an acacia had caught his attention. I was too slow to see the bird. Treecreepers try to escape observation and predation, Alec explained, by climbing around to the opposite side of trunks.

Alec always carries a magnifying glass when he goes bushwalking. He talked about the need to understand natural patterns close up. When his children, nephews and nieces were young, Alec gave them a sense of familiarity with Pioneer Park bushland, its animals and insects. People, he told them, are part of wider natural processes. Ecological ties are especially visible and active in diverse forest communities. Alec described to his young family members how eucalypts and wattles hold carbon and release oxygen. Responsible engagement with places depends on ecological consciousness, Alec believes, on personal awareness of natural interactivity. He wanted his children and junior relations to understand their physical connections to land. When they became adults, he told them, they could influence the wellbeing of places like Pioneer Park. Again as he spoke to me, Alec stopped mid-sentence. He pointed into a wattle tree. Above our heads, a branch held a fantail nest woven from stringybark fibres and grass.

As we returned up the slope towards the car park, Alec told me another story about Pioneer Park. For much of his adult life, Alec worked as an electrical linesman with the Northern Riverina County Council. He relished the freedom the job gave him to encounter and engage with varied places across the region. Climbing an electricity pole one day in the early 1960s, he lost his footing and fell. He landed flat on his back. While he escaped major breakages, his collision with the ground induced a debilitating condition. He began to suffer intense and lasting migraine headaches. To find relief, he walked into the forested hills of Pioneer Park.

Alec sat under eucalypts, watching and listening to the activity of woodland life. One time, he saw a willie-wagtail build a nest of spider webs and feathers. As he passed restful hours on the shaded and lively hillsides, his painful headaches lifted and vanished.

~ ~ ~

Few trees grew on the stony hill northwest of Cootamundra when my parents bought our small farm in the late 1970s. Just a scattering of mature stringybarks, red gums and bundy box, a single kurrajong, and several drooping she-oaks, ancient and broken. Grey stumps scarred long ago by ringbarkers rose above summer swathes of wild oats, devil's claw, ripgut brome and Paterson's curse.

'We were such fools', our neighbour Joe Manning told me. Ringbarking teams worked the hill generations before Joe cleared the granite rise of dead trees and forest regrowth in the 1950s. Joe led a draught horse up the slope to pull the weathered trunks down. A fire established, he shovelled coals between stumps and fallen timber. Joe ringbarked the saplings. He poured poison from an old kettle into the axe wounds of eucalypts to ensure the young trees wouldn't sucker.

Stock loved the stony hill, Joe remembers. It was 'sweet country' – sheep and cattle found grasses and forbs growing amid the rocks especially palatable. More pasture grew with the trees gone. In nearby paddocks, Joe cleared other scrubby and regenerating stands. New, powerful machines made clearing an easy task. Across the district, kangaroo grass and other native forage plants vanished as landholders spread phosphate fertilisers and applied more intensive farming techniques. Regretfully, Joe noticed unforeseen consequences. Noisy flocks of green budgerigars disappeared. Curlews no longer wailed at night.

Where the slope of the granite hill eases, on the southeast side, a bare expanse of salty earth appears to be spreading. Without a diverse mix of trees, shrubs and grasses, the hill no longer keeps rainfall from watertables below.

A few years ago, we fenced the stony slopes and hilltop from livestock. Three rows of barbed wire above two strands of plain give kangaroos and wallabies easy access. The winter after the fencers enclosed the hill, people came to help plant trees and shrubs, to camp on a sheltered flat and celebrate the work. We spent two days finalising our preparations for the event. From a revegetation nursery near Bethungra we bought plants propagated from local stock – violet kunzea, prickly tea-tree, narrow-leafed hopbush, western silver wattle – adding to our collection of bundy box, early wattles and drooping she-oaks grown from seed collected on the farm. We bought thick sausages and chops for the barbecue, split green peas and bacon bones for soup. I revved up the chainsaw and cut heavy lengths of red gum for campsite fires.

A cold, gusty change swept the southwest slopes the night before people arrived. The Bureau of Meteorology issued a sheep graziers' alert and predicted snow on the mountains, even along the tablelands. I climbed the windswept hill in the morning. Cold rain blew across the slope. Swinging a hoe, I cleared patches in the wet surface of grasses and weeds – hundreds of sites for young plants. I put a seedling guard beside each circle of shaved earth, holding with loose stones the plastic sleeves from the wind. Next morning people would lower the guards around bamboo stakes hammered into the stony earth beside each plant, offering seedlings protection from rabbits, hares and scorching summer winds.

The previous spring we had planted hundreds of *Acacia deanei*, dark green and feathery-leafed, beside the gully at the base of the hill, inside another area fenced to exclude livestock. Many had died when summer came early. After preparing the planting sites across the wet and wintry hill, I walked down to the watercourse later in the afternoon. I carried a wheat bag to collect plastic guards and bamboo stakes from the dead wattles, for reuse uphill the next day. I passed a place where I once found a hand tool beside a cattle track, a pale flake struck from fine-grained stone, narrow and sharp. As a child I spent days down here, unearthing old domestic rubbish somehow entombed in the gully wall. I found glass bottle tops, roughly turned ceramic inkwells and a rusted saucepan holding a peach stone. On the nearby hillside, I once found a sulky step on top of a stone pile, geometric designs stamped into its footplate.

In search of dead wattles and surplus guards, I reached the edge of a broad, deep basin carved by churning water, a legacy of violent, forgotten storms. A green plastic sleeve lay inside the hollow, lifted and carried from bamboo stakes by a brisk wind. I negotiated the steep bank, stepped across the gully floor, and shoved the guard into my wheat bag. An old stringybark tree fell into the watercourse here years ago, as soil around its roots washed away. One branch pointing skyward grew tall and sturdy. I paused. Other mature stringybarks – wide trunks dark in the evening gloom – stood quietly above the sheltered depression.

My lips burned, swollen by hours of exposure to cold, biting wind. I felt the slopes of the stony hill in my legs and the weight of red gum branches in my arms. Clouds glowed purple in the darkening sky. Time, place, stringybarks, growing wattles, dead wattles, all seemed to whirl and blend before me and through me, as I stood in the earthy basin, tired and hopeful.

Jo and Graham arrived after nightfall with Roi and Mareve, Israeli backpackers working for a few days on Graham's farm near Narrandera. We unloaded five hundred seedlings – drooping she-oak and stringybark – thriving in black plastic trays. Graham grew the young plants from seed collected on the dark hillside beyond the house, where cold wind whispered in the crowns of elderly trees. In the morning more people came, parked beside the fence, and walked up the slope to help plant. Rain clouds had vanished overnight but the firm westerly kept blowing. In the clear winter air we planted on the eastern slope, angled to the morning sun and sheltered from the wind.

Hazel brought Mohammad and Assad, Afghani refugees. Like other members of the Hazara minority group, the young men had suffered persecution in Afghanistan under the brutal Taliban regime. They fled to Australia, arriving by boat on the northern shore and claiming asylum. The federal government detained them in a detention camp at Woomera in the South Australian desert. After months of waiting, immigration officials gave them temporary protection. The two men, skilled tailors, found work in a Cootamundra sheepskin factory making boots.

Mohammad seemed cautious as we ate lunch, thinking. Jim complemented him on his English, and asked how the language classes were going at the tech college. Wiradjuri language lessons had recently started in Cootamundra, I mentioned. It appeared Mohammad didn't know that a cultural group other than European settlers had a presence in the region. My mother tried to explain. 'There were Aborigines here, then they were cleared off the land', she said, awkwardly. Across the wide gulfs of difference separating Assad and Mohammad from my people and myself, no proud or comforting settler myths held ground and gave shelter. The two young men knew well the trauma of dispossession and violence. They sat silently.

Afterwards, back on the hill, Melissa heaved a tall iron crowbar to the sky and it fell with her weight into damp stony earth, the dull sound of metal on stones held loosely underground. She repeated the action, levering soil and granite rocks aside to make space for a stringybark sapling glowing rosy and green at her feet. Roi and Mareve worked away, pressing dark earth around the naked roots of a juvenile bundy box, threading bamboo stakes through a plastic guard.

Nearby, rock ferns and lomandra poked out from stony crevices into winter sunshine. For so long, sheep and cattle had kept the plants trimmed back. After the fence went up and livestock were excluded, shoots and leaves grew too from the bases and lower branches of elderly red gum, stringybark and bundy box trees. On the eastern side, where Joe Manning and earlier ringbarkers had left a clump of trees, limbs running close to the ground were now bushy with growth. Fresh bursts of blue foliage closed the gap between upper branches and the ground, a space formerly kept nibbled open by livestock. The new growth offered animals and plants rare shelter from winter and summer winds.

Mareve positions bamboo stakes and a plastic guard around a bundy box seedling on Treetops, Cootamundra district.

At the end of the short winter day, as the sun dipped and lost warmth, Roi and I walked over the hilltop to retrieve tools left behind. Roi told me about his tiny garden of fruit trees beside his home in Israel. He swept his arms out and said how lucky my family was to have all this space. I tried to say something of the history that led us here. I turned and pointed south towards Mount Ulandra, its forested bulk darkening as night came, and said I knew a story of killings there. On Yammatree station beside the mountain, settlers had executed Aborigines for 'stealing' livestock, a descendent of Yammatree's pioneering family once told me.

The sun vanished behind a line of hills studded with elderly red gums. Cold wind blew on. 'Come back one day Roi', I said, 'to see how the plants are growing.' Roi laughed. Earlier in the day we'd talked about his year of compulsory military training. He and Mareve were returning to Israel soon. Over recent months, news reporters had told horrific stories of suicide bombers and terror, a relentless and escalating war. A tall grey stump beside us bore a ringbarking scar around its girth. Butchered limbs reached out. Nearby, a plastic guard gave shelter to a drooping she-oak seedling, carefully watered, its tender white roots feeling the earth.

Months later, the spring weather was hot and dry. Drought again descended. As stringybark saplings rose meekly above guards, fierce westerlies curled soft leaves brown and crisp. Hungry kangaroos pushed the green sleeves down to nibble she-oak leaves. According to botanist Joseph Maiden, thirsty settlers chewed the dark green strands. Acid in the drooping she-oak leaves generates saliva and relieves thirst.[1] Perhaps marsupials find the same. Once a month we watered the dwindling population of young trees and shrubs. Weighed down by a fire-fighting tank filled with water, the ute made firm tracks through the blond expanse of wild oats and red grass. Buckets spilled as we stepped over rocks towards seedlings. One night I dreamt of rain, the stony hill cloaked in green, wet grass.

Ossie Ingram was born at Erambie mission beside the Lachlan River at Cowra just after World War I, and grew up in the Narrandera district.[2] In the 1980s, the Wiradjuri elder told a story about Muri, 'a maiden of the mist' who fell in love with the north wind.[3] As the wind blew from the north, the story goes, rain clouds gathered and broke, filling waterholes and offering Muri abundant moisture. Then winter approached. A cold wind started to blow from the south, competing with the north wind for Muri's attention. The bitter southerly blew and blew, blocking the northerly wind, preventing rain, and banishing warmth and life. As moisture disappeared from billabongs and the land, Muri turned cold and died.

I thought of the story told by Ossie the dry and dusty February after our planting, when a firm, warm northerly swept the southwest slopes and southern tablelands for three days. Weather charts showed an intense low-pressure system approaching from the west. Circular isobars traced the clockwise path of winds. A large body of unusually moist air channelled down from tropical waters across the red sand dunes and spinifex plains of central Australia. I visited the Bureau of Meteorology website, and found the current rainfall map generated by radar from Wagga. Wide, mottled bands of grey, blue and yellow drifted eastwards, signifying copious, gentle rain.

The following spring, Penny and I drove south to a planting weekend near Oura village, beside the Murrumbidgee River upstream from Wagga. Here, elderly river red gums rise from a floodplain area reserved for recreation and travelling stock. Janu needed long ropes to bind floodlights to the girths of ancient eucalypts at the campsite. After nightfall, branches and leaves glowed lime green and red.

Weeks before, an Oura resident had driven a tractor with a ripping blade along parts of the reserve selected for planting. The device had torn furrows into grey floodplain earth. In spring sunshine, we followed the lines across the uneven plain, planting thousands of river oak, bottlebrush and red gum saplings. It was easy to imagine floodwater lapping at eucalypt trunks, receding through fallen branches and grass tussocks as frogs called and waterbirds foraged for yabbies – a regenerative event denied by dam walls of concrete and steel in steep river gorges upstream.

Graham wore a T-shirt designed by April, inscribed 'Community,

Regeneration, Celebration'. April talked about connecting with elements of nature sure to remain active far beyond human life spans, and celebrating those bonds with dance:

> To be taken out of working at a desk during the day and going and planting a tree and thinking this is going to grow, hopefully, and be as big as these other trees. It's the interaction with the earth, and then you come and you dance.

Jo spoke of a need for time and quiet to listen as people have for so long here. 'Just that Aboriginal idea of the land constantly singing a song to you to explain how it works and how you can operate in harmony with it', she said. 'You just have to listen to it, learn how to listen to it.'

As evening came, red gum branches burning on the riverbank gave warmth and light. Overhead, pairs of wood ducks and noisy cockatoos returned to nests and roosts. Janu fed Australian birdcalls from an old vinyl record into the mix of electronic trance music. Beside the fire a cattle dog sat up, ears pricked, as speakers cast willie wagtail chatter, wood duck calls, and the aching cry of a crow across the darkening floodplain.

Conversing with place

Ardrossan is a harbour town on the rocky western coast of Scotland. The placename 'Ardrossan', according to one local source, means 'a rocky height, terminating at a rocky point in the sea'.[4] Shipbuilding and sea trade flourished at Ardrossan in the nineteenth century. When my father and mother visited late in the twentieth century they found a bleak and depressed town. Manufacturing and export companies were gone, the commercial port closed.

I downloaded a copy of an old watercolour painting of Ardrossan harbour from a website. In the tranquil scene, masts rise from ships resting in glassy water. Seabirds fly above a sandy beach. Rugged mountains fall to the sea.

My grandfather called his grazing property west of

Cootamundra 'Ardrossan'. Soon after marrying in 1940, he and my grandmother moved into a new home of white cypress pine felled in a nearby paddock. A generation later, my parents settled into the weatherboard house on Ardrossan, sheltered by remnant bushland on a gentle hill, my first home. Shapes and particularities imprinted themselves deeply. I remember a gravelly dam wall, a hillside of bleached grass, cypress pines against blue sky, yellow box trees beside the track, kurrajongs my mother planted at the ramp.

My grandfather never visited Scotland or the distant town of Ardrossan. His family lived there for several generations in the nineteenth century, perhaps when someone painted the picturesque seaside image I found on the web. Over the course of a century, his paternal ancestors moved west to Ardrossan from the eastern coast of Scotland, to Australia and to southern England.

Modern restlessness and desire for material and social advance drove my ancestors and other western Europeans to distant places, then on again. Colonists showed commitment to movement and progress, not places or communities. They built transportable, patrilineal identities. Unlike active relationships with people and land, family histories and trees could be written down, charted, memorised, taken away. The name my grandfather selected for his grazing property on the southwest slopes didn't reflect natural particularities of the local area. His choice, 'Ardrossan', spoke of movement down patrilineal and geographic pathways. The name arose elsewhere, like the demands of distant markets for the wool he grew and sent away.

My sister and I notice the air when we visit places near Ardrossan, the light. More than a sentimental yearning for childhood terrain, perhaps our responses and desires are innate and functional. Here, it seems land began drawing us into ties of mutual care, into lively obligations and a sensual existence.

Seven years old, in the garden at Ardrossan, I asked my father if he would cry when we left the property and moved closer to

Cootamundra. We drove past Ardrossan recently, now owned and worked by someone else. He wept quietly in the passenger seat and talked of old hopes and plans, unrealised.

Months later, one autumn afternoon, my father and I returned, stopping the car beside a back paddock of Ardrossan. An elderly grey box tree threw shade across the uneven ground and tussocks of native grass between the boundary fence and the bitumen. On the other side of the wire netting, open farming paddocks stretched north and west. I held my tape recorder. My father tried to describe his recurring dream of the place before us. He grasped for words to capture images shifting at the back of his mind:

> In my dream the trees seemed to be taller and thicker, and possibly not so long dead. But very noticeably with areas throughout the timbered area where there were carved trees, or buildings built out of the trees, or in the trees. There were things built on the trees, too, I suppose, remnants of boards being fastened on trees in a significant sort of a way. And certainly carvings in the trees – but, my memory is more sort of European-type things than Aboriginal-type things, and a quite weird sort of a feeling. Because even though I was there in them, they sort of weren't there, in a way. Sort of the hidden corner you walk around and there's a memory there you didn't think was there.

Four decades ago, a ringbarked forest of yellow and grey box trees cloaked these paddocks beside the road. My father remembers broken trunks rising from rough ground. Ringbarkers killed the eucalypts generations before, when Ardrossan was part of Retreat station, a grazing property where my grandfather was born and raised. Corriedale sheep grazed native grasses and forbs among rotting trunks and limbs. Workmen regularly collected truckloads of the weathered timber. Beside the homestead on Retreat, the men spent days cutting firewood with a circular saw driven by a tractor. Fertilising contractors bounced old trucks across the uneven paddock, spreading superphosphate over native clover and grasses.

The wide stand of decaying trees on Ardrossan embarrassed my father's family. By the 1960s, most landholders in the Cootamundra and Temora districts had cleared remaining patches of dead timber and sown cereal crops and imported pasture species – they had 'cleaned up their country', my father said. He organised for a neighbour to help. They fastened a bulky chain between two tractors. Dead trees fell as the diesel machines strained in tandem. The men used the tractors to drag and push the brittle wood into heaps for burning. My father remembers a feeling of satisfaction and accomplishment as he ploughed the paddock for the first time: 'I'm building something here, I'm developing something. This is good, this hasn't been done before, this is the way forward.'

Creating space for industrial futures requires erasure of particularities and histories. Machinery groomed the paddock. Weathered logs, native grass tussocks and stump holes vanished. Ploughs and crop-seeding equipment destroyed a series of mounds my grandfather called 'gins' graves'. The earthen humps were probably natural formations, my father thinks, but he's not sure. His family never talked about local history.

I asked my father to describe the feelings underlying his recurring dream of the paddocks extending before us. He spoke about lost mystery and the severing of connections:

> A sense of loss, I think. Maybe just a sense of nostalgic loss. No, I think it was a bit more than that. There was a loss of something of significance. And I think it possibly did relate to those trees. Because there was some link with them that was certainly gone when we cleared them. A link with another reality, a past, a thing that was long past. There was a visible, tangible link. The trees were evocative of things past, but things not particularly clearly identified.

Something more than dead trees and uneven terrain disappeared from the land my father and our neighbour cleared. Maybe a sense of vanished potential for dialogue and connectivity keeps evoking the haunting dream. Tractors and ploughs erased intricate patterns – cultural and natural. Burying native clover and grass, constructing blank, industrial space, the action eliminated opportunities to comprehend and work inside a dynamic and expressive local ecology. Possibilities for new, mutually nourishing relationships between land and people evaporated. When my family sold the farm almost thirty years ago, we lost any chance of finding alternative ways to engage with the paddocks of Ardrossan.

Later, my father remembered a second recurring dream of the same place. Moist and alive, the woodland in this dream offers life and security. Painful feelings infuse the dream:

> In my dream I am walking through the area, realising we have sold it and that I am trespassing. The area is timbered again, though in my dream I remember it as always being timbered. The trees are mature and there are understorey plants. I don't know in my dream what types of trees and plants but I have no curiosity about this. It is always wet though not raining. The timber is not dense but there are enough sight breaks and clumps of trees to make me feel reasonably secure from being seen. It is a fecund area. But no people, except me, only plants. There is a sense of loss because I know I will have to leave soon.

~ ~ ~

Just west of Birrego, a farming district south of Narrandera and the Murrumbidgee River, gentle hillsides ease into the Riverina plains. Here, on the dry margins of the southwest slopes, sparse columns of white cypress pines and dark green globes of kurrajong trees rise from paddocks of wheat and sheep. A small timber church marks the centre of the Birrego district. Mosaics of wattles, native grasses and eucalypts grow along road reserves.

The sparse foliage of hooked needlewood, *Hakea tephrosperma*, gave little shelter from a cold southerly wind arriving one November morning, slamming a door and waking the house. Over breakfast the radio predicted storms. Winter rains and spring warmth had nourished Birrego district crops, pastures and roadside grassy woodland remnants. Crops and native plants stood heavy with grain. Hakea flowers resembling creamy spiders decorated the small colony of trees Penny and I harvested. 'Turn right at the church, and they're about three quarters of the way down the road, over the rise', Graham told us after breakfast, before driving away in a wheat-laden truck towards the Boree Creek railway silos. We arrived at the needlewoods with gloves, a ladder, secateurs, buckets and a woolpack.

Trying to avoid the spiny hakea leaves, we reached into the foliage and twisted the woody seedpods away. Some pods had already offered their winged seeds to the breeze. I imagined the sturdy capsules we were harvesting opening days later inside the woolpack left in the sun. And soon, the seeds sprouting in plastic trays. When autumn rains came, Graham and other Landcare group members would plant the hakea seedlings alongside wattles and eucalypts, into revegetation belts across farmland. Small birds nest amid the protective spiked leaves, and eat pest insects on paddock trees, crops and pastures.

The hooked needlewoods resisted our efforts. 'Have a look at this', Penny said. She turned her forearms, pricked red by the sharp, curved points at the tip of each needle-like grey leaf.

Two thick, curving trunks rise above the young, sturdy plants we harvested. Coloured patches of lichen decorate the furrowed bark of the venerable hakeas. I sensed an obligation to acknowledge the elderly pair of needlewood trees, to explain our hopes and actions, and to ask for permission to harvest. Perhaps we were already performing respectful rituals, at Birrego and elsewhere: travelling long distances, the repetitive harvesting motions, daily watering seedlings in pots and trays, careful planting into prepared earth.

Grey clouds gathered to black as the cold southerly blew. We heard machinery in a nearby paddock, where the driver of a combine harvester rushed to collect a crop of barley before the storm broke. Hail and sharp downpours flatten modern cereal crops – millions of brittle stalks supporting heavy grains swollen by fertiliser. Unlike crop species grown on the southwest slopes, needlewood hakeas and other grassy woodland plants generally survive storms, drought, frost and fire.

As the storm clouds gathered above Birrego paddocks, I noticed hakea branches holding both ripe seedpods and new flowers protected by spiked leaves. Belting hail might dislodge the blossoms, but never the woody capsules. Open foliage and forked limbs capture and channel rainfall down furrowed trunks into an absorbent groundcover of moss. Even if we greedily harvested every mature seedpod, summer rain would nourish seeds forming inside small green pods along each branch.

Old kurrajong trees stand in paddocks beyond the roadside hakea trees. Generations of wheat farming and sheep grazing ensured that no saplings emerged in paddocks. Native shrubs and grasses able to conserve moisture and hold soil vanished from farmland. Dead trees and salt scalds on farmland beside the road to Narrandera are obvious consequences of dramatic alterations imposed by agricultural development.

Apart from time spent at boarding school and university in Melbourne, Graham Strong has always lived here at Birrego, on farms owned by his family. The young farmer described to me the 'uncomfortable feelings' he experiences 'standing next to a four hundred year old kurrajong tree in a bare paddock'.

We talked further about his encounters with elderly, isolated kurrajongs on farmland. He built a sturdy fence around one large kurrajong, enclosing a wide area around the tree. 'That was to protect it', Graham explained,

> and to give it some respect and try and restore some of that dialogue, to say 'I am listening to you.' I know it's not fantastic, but at least this is a start, this fence around this kurrajong. 'At least shield you from some of the impact, at least get some shrubs growing under you so birds can come in and eat insects that are

eating your leaves.' That's an initial step I can take towards establishing dialogue, and cooperation, and respect, mutual respect. 'You give me shade, you give me this beautiful tree, I can give you this back.'

Graham doesn't embrace the discourse of 'natural resource management' and 'sustainable agriculture', the rhetoric of strategies designed to maintain industrial, productionist agriculture. Once, Graham and I talked about the alternative concept of 'regenerative agriculture'. Unlike 'sustainable', the label 'regenerative' acknowledges a painful history of suppression, fragmentation and disorder. Connectivity is acknowledged and nurtured. Rejecting the term 'sustainable' and adopting 'regenerative' invokes formidable new relationships.

Opening dialogue with a subjugated and wounded entity is not an easy, comfortable process. Graham wrote:

> I sometimes feel equally awkward or uncomfortable to use the term 'Regenerative Agriculture' because it poses such a challenge to my soul and deeply confronts me and challenges me to feel, think and act deeply. The ethical obligation emerges like a tortured soul begging for recognition, love & nurturing.

Ecological fragmentation and loss brought by agricultural development and monocultural production make dialogue and learning difficult. Extinction of grassy woodland species and the simplification of natural processes block understanding of the dynamic potential of local ecologies. When conversations between people and land begin, dialogue is unavoidably erratic and halting. Industrial development has destabilised local ecologies across the southwest slopes, undermining capacity to comprehend natural patterns. Monological relations of power further disable learning.

Nurturing rural places, and allowing rural places to nurture and enliven us, requires careful attention to the intricate patterns and particular needs of local ecologies. Dialogical relations arise when the diverse expressiveness of places and other species is acknowledged and sought. At Birrego, Graham Strong and his family are developing farming systems responsive to the particular histories and life of local terrains. Recognition of ecological and social fragmentation wrought by colonisation underlies their desire to restore biological diversity and natural integrity.

Histories of Wiradjuri dispossession remain active at Birrego, informing decisions Graham makes. In the middle of the 1990s Graham helped form a local Landcare group. He invited three Wiradjuri elders to the inaugural meeting. Graham remembers himself speaking at length. 'I wanted to show that we respected their law', he told me, 'not that we even really knew what their law was, and that we were trying to do the right thing.' It was an awkward moment. The elders remained silent.

On a separate occasion, an Aboriginal man from Narrandera came to Graham's homestead to ask for access to the farm. The visitor wanted to collect a particular grub, excellent fishing bait, which burrows into the dead trunks and fallen branches of kurrajong trees. Graham readily gave permission, but felt uncomfortable doing so. Powerful settler notions of land ownership and control seemed to lack legitimacy. 'You feel that clash of laws', he said. Unexpectedly, Graham felt nervous speaking to the visitor. He saw the Wiradjuri man as 'a policeman', someone perhaps making a judgment based on Aboriginal law. Graham yearned for dialogue:

> I felt like talking about all the ways we're trying to look after the land, you know, regenerate the trees and the natural world, and I felt I had to justify myself, that I was trying to fit in with at least elements of, through my limited understanding of Aboriginal culture, of tribal law.

Graham and I drove across the land he farms with his parents. Paddocks we crossed looked different from neighbouring ones. As well as lucerne pastures and wheat crops, the Strong family has planted entire wheat paddocks to old man saltbush and native grasses. Graham sows wide bands of acacias with the same broadacre seeding equipment he uses to plant cereal and canola crops.

After one pass of the wide machinery, with good rain, tens of thousands of wattle seeds collected on local roadsides sprout and grow in moist red earth. Graham plants silver banksias, quandongs, eucalypts and other local species among the wattles. 'Now the land is really coming alive', he told me.

Only five years after the start of the revegetation work, the dynamic expanses of native vegetation had halved the volume of expensive insecticides needed to stop pests damaging pastures and crops. The growing native plants harboured many insects and small birds that devoured troublesome bugs and caterpillars.

Graham pointed to several young kurrajongs rising inside a revegetation belt. The trees are fire retardants, he explained. Kurrajongs retain moisture throughout summer, and their dense foliage helps block wind and flame. Similarly, as sheep graze deep-rooted saltbush, the woody plants keep producing fresh, succulent growth in dry months. Moisture drawn from subsoils into green leaves and branches builds fire resistance.

Between the saltbush rows, Graham is establishing swathes of summer-active forage plants like kangaroo grass. He hopes to avoid what he describes as a 'dead system' prevailing across the southwest slopes between October and April each year, when paddocks cloaked in exotic annual crop and pasture plants turn dry and bare. Without deep-rooted, summer-active perennials, soils lose so much moisture that even microbial life vanishes. Farmland becomes static, vulnerable to fire and erosion. The absence of perennials able to respond to summer rain allows moisture to soak down into watertables, exacerbating dryland salinisation.

In one paddock Graham showed me kangaroo grass tussocks green and growing, despite especially hot, droughty conditions. If a bushfire approached, he planned to gather his sheep inside relatively moist and fire resistant saltbush stands.

Graham accepts fire as part of the local ecology. Communities of plants native to the southwest slopes, he explained, are adapted to fire. Working with the patterns and forces of the land, planting kurrajongs, saltbush and kangaroo grass, reduces the chance of catastrophic bushfires. Windswept paddocks cloaked in dry, lifeless crops and pastures across the region in summertime invite and often deliver fiery disaster.

We stopped beside a solar panel, angled to the sun. Graham welded a stand for the panel from an old plough disc and a segment of pipe. Coloured plastic clips and coated wires linked the panel to a two-strand fence. Electrified wires bisected a paddock of saltbush and native grasses. Graham regularly transfers mobs of sheep into paddock segments bounded by the movable electric fences. Sheep intensively graze small areas for a week or a few days, depending on seasonal conditions and pasture availability. The mob is then shifted to another paddock segment, where over past months woody shrubs and perennial grasses have grown freely.

As the grazed saltbush and native grass portions hold sheep for only short periods, roots beneath each perennial plant remain active to a great depth. Denuded shrubs and grasses respond vigorously, a capacity lost when plants are continuously grazed and only shallow roots survive.

Kangaroo and wallaby grass, old man saltbush and other native perennials are closely adapted to the Birrego district. Deep roots, water-holding capacity and hardiness enable the shrubs and grasses to grow fresh foliage throughout the harshest summers. Watertables are lowered, the chance of dryland salinisation reduced. In wintertime, as cold winds sweep the southwest slopes, the native plants offer sheep warming feed and shelter.

Wendell Berry considers the modern, industrial model of agriculture monological in method and arrogant in character.[5] Careful, responsive attention towards natural particularities is blocked. Never has the industrialised farming system, Berry writes, 'asked for anything, or waited to hear any response.' The dominant mindset of industrial agriculture identifies production and profit maximisation as primary goals and standards of measure. Supportive alliances with natural forces are not considered.

An alternative style of farming, one instead taking the well-being of people and land as central targets, 'would approach the world in the manner of a conversationalist', Wendell Berry suggests. Seeking dialogue and intimacy with rural places, an ecological model of agriculture would abandon the future-oriented, powerful strategies applied across the farmlands of western nations. People rejecting the established model 'would undertake to know responsibly where they are', writes Berry.

> They would ask what nature would permit them to do there, and what they could do there with the least harm to the place and to their natural and human neighbours. And they would ask what nature would *help* them to do there. And after each asking, knowing that nature will respond, they would attend carefully to her response. The use of the place would necessarily change, and the response of the place to that use would necessarily change the user. The conversation itself would thus assume a kind of creaturely life, binding the place and its inhabitants together, changing and growing to no end, no final accomplishment, that can be conceived or foreseen.

Inside Birrego district paddocks, Graham Strong and his family converse with place. Farmland offers understandings of connections and flows. As soaking rains draw nutrients beneath the shallow roots of cereal crops, the much deeper roots of saltbush and kangaroo grass return leached elements to the surface. Sheep fed on perennial shrubs and grasses deposit organic matter and minerals across adjacent cropping paddocks. Graham is always 'catching up with the land', he explained, trying to sense and respond to the immediate needs of the local ecology, to fold primary production into complex and shifting patterns.

Fostering natural productivity and bolstering the resilience of land requires listening and responding to rural places. 'Listening' involves broad sensual engagement – feeling moisture in leaves and earth, looking at colours and textures, listening to animal calls, sensing the direction and temperature of wind.

'If I were a sheep, how would I see this paddock?' Graham asks himself. Low voltage electric fences enable better understanding of what sheep desire, as the animals are able to push through the wires. Semi-permeable, such fences are instruments of dialogue, not control. Often sheep temporarily leave the enclosure to graze young wattles and saplings inside a band of regenerating grassy woodland. Graham knows it's time to shift the mob if the sheep don't return, or if the animals are pushing through fences into areas of fresh saltbush. Seen as integrated components of a living system, sheep become expressive agents of place.

When we returned to Graham's homestead from a paddock tour, I noticed old man saltbush lining the front garden path. Kangaroo grass and golden wattle grew inside the garden fence, alongside the weatherboard house. The homestead and its garden seemed part of the flourishing life of paddocks and gentle hills beyond. Here at Birrego, as seasons pass, conversation draws land and people closer together. I felt hopeful speaking with Graham. Fragmentation and disorder, it seemed, need not be enduring legacies of colonisation. Perhaps careful listening to places and history, respectful dialogue with people and other species, can bring regeneration.

epilogue

'Straight lines would look really ludicrous', Owen said as we stood in the paddock between the creek and a steep range of forested hills.

A curved line of turned earth and seed lay at our feet. Owen pointed across the flat towards other wide spirals scraped into the ground, separated from each other by grassy expanses. Owen had worked outwards with a direct seeder hitched to a ute, sowing thousands of eucalypts, wattles, and understorey plants in widening arcs.

After a long, hot summer, most of the seedlings had failed to grow. Owen planned to try again if the drought lifted in coming months, retracing the same path. 'I want to try and break up this bare landscape and get some structure', he explained.

We walked towards the hills. Owen stopped beside a low mound, the detritus ring of an enormous, vanished tree. For centuries here, leaves and bark fell from the branches of a mature eucalypt. Bulky roots swelled the ground. Windblown earth and humus built up around the trunk during storms.

Owen pointed to a dip in the centre of the mound. Someone in his family, he explained, had burned away the stump of the ring-barked tree. A long depression snakes out from the detritus ring, where fire consumed a major root. Across the paddock at regular intervals were other mounds, evidence of an open forest of established box trees amid grasses and forbs. 'They were sitting ducks', Owen said.

Hugh and Catherine Whitaker settled east of Junee beside the upper reaches of Mitta Mitta Creek in the 1870s. Some branches of the family stayed in the district, others moved away. For most of his childhood, Owen Whitaker lived on a farm southwest of Wagga on the Bullenbong Plain.

In the 1960s, Owen's father inherited Oakley, a small property in the Mitta Mitta valley, from a great-aunt. His parents decided to return to the district where the Whitaker family had lived for almost a century. From his uncle they bought a larger and more intensively managed farm, Kimvale, down the valley from Oakley, and ran the two properties together. A decade later, Owen took over the farms and his parents moved away. He lived with his wife and children on Kimvale.

One autumn afternoon I walked with Owen across Oakley, the smaller and less transformed of the two family holdings. Owen loved visiting Oakley as a child. 'It was always a bit of a wild, different, funny place to our other production farm', he explained. Unlike Kimvale, where land was altered to suit industrial methods of production, natural patterns prevailed on Oakley:

> We'd come up here and it would be like bush. It's not true bush as we know it now. But it was always different. And it had these old ramshackle buildings on it, and always had heaps of grass, scrub and trees and rabbits and kangaroos, and it was just so totally different to what we were used to.

We sat and talked on the rise above the creek, beside a rusting tank and the remains of a shearing shed destroyed in a bushfire. As we ate lunch, a peregrine falcon swept through the air before us. Owen pointed to a gorge in the hills where pairs of the predatory birds nest among rocks. Once when he walked near there, a falcon flew down and challenged his presence with a high-pitched cry. Sometimes at Oakley, Owen encounters wedge-tailed eagles:

> They come down for a look and give you a bit of an eyeful and give you the distinct feeling that you're a bit of an outsider. And we're talking about the power of landscapes and country and it's that

> same sort of thing. I think what attracts people to certain landscapes is that it is wild, it's untamed, there's a power there. Yeah, it's still its own entity.

Coastlines, mountains, rainforests, and deserts tend to draw tourists and inspire imaginations today, not agricultural regions. Farmland supplies food, fibre, and export dollars. Deeper meanings are evoked elsewhere. 'People want to go away and be inspired by other landscapes than these production landscapes', Owen said,

> but is that because they never really were inspiring, or because they've lost what it was that spoke to people, that touched them, that elevated them, that lifted their spirits, that rose above them?

Rural places have lost power to attract and inspire: 'I think this landscape's a bit tired and it's sad. And if people go away they don't want to go somewhere that's sad.'

Owen talked of 'the silent bush', scrubby areas and remnants of grassy woodland now devoid of animals other than birds. When his elderly relations were young, 'it was the other animals and noises and beings in the landscape that made it really special'.

His grandfather once had a pet koala. As he relaxed inside by the fuel stove at nightfall, the creature clung to his leg. Local and urban people found delight in the lively and diverse nature of the Mitta Mitta district:

> My family would talk about the good times they had in this landscape. People would come down from Sydney. And, you know, they were very much envied. They used to send game and food up on the train in crates. They'd work hard all week. They'd go and play cricket or tennis on Saturday. Sunday they'd go to church, and then after church there'd be people coming to visit. And some Sunday afternoons there'd be a fox drive. All the blokes would go out on horseback with their shotguns, and that was a community thing. They obviously loved interacting in the landscape. Yeah they went to the coast or down to Sydney very occasionally, but there was still a lot in the landscape to cherish.

As Owen farmed his two properties – Oakley, the less altered property higher in the valley, and Kimvale, the larger and more industrialised farm – he began to observe striking differences. Down the valley on Kimvale, stock often suffered from nutrient deficiencies and disease. Decades of high-input, high-output farming methods had depleted soils of essential trace elements, nutrients and organisms. Pastures no longer gave stock a balanced diet, and intensive farming was causing soil acidification.

In contrast, sheep and cattle thrived at Oakley, grazing a wide range of plants and drinking fresh spring water from the creek. Unlike on Kimvale, Owen spread little fertiliser across Oakley. He regularly top-dressed only a small area of sown pasture. Nor did the diverse stands of native grasses and forbs need insecticides or herbicides to stay healthy and productive. Soils remained stable, and showed no signs of acidification. Owen started listening to the land:

> I started to learn a lot. And I started to think: 'Well, here's a great insight, in that there's totally different forms of country, and they're telling me different things.'

When Owen considered the minimal amount of inputs and time needed to maintain the productivity of Oakley's paddocks, he realised the small property offered a far greater monetary return per hectare than Kimvale.

Walking through a paddock, we stopped beside a low shrub where a quail or groundlark had recently nested in the grass. White pellets of bird manure lay inside a faint depression. Here was 'a hotspot of nutrient', Owen explained, extremely high in nitrogen and phosphorus. When a bird made a nest in the shelter of a shrub, nutrients essential for plant health and productivity soaked into the earth. 'Too slow for our reckoning, or our contemplation, but sure and stable and happening nonetheless', Owen said. Without the shrubs and various other plants, such interactive processes can't happen. Natural productivity and ecological stability are lost.

epilogue (255)

Across the southwest slopes, where settlers replaced diverse and dynamic communities of grassy woodland with a tightly managed network of paddocks comprising few plant and animal species, ecological disorder will continue to mount: 'As soon as you really modify a landscape beyond a certain point, you're only on borrowed time.'

Owen Whitaker,
Oakley, Mitta Mitta district.

Bringing ecological stability and wellbeing back to farmland requires the recognition and disabling of destructive economic dynamics. The imperatives of financial institutions and globalised markets ensure that farmers are rarely able to respond adequately to the shifting needs of local ecologies. As economic pressures tightened and new technologies emerged after World War II, more intensive methods of production undermined the natural productivity, resilience and stability of farmland. Generations ago, Owen explained, farmers enjoyed greater degrees of independence and management flexibility. He described the historical situation and the changes needed today for ecological and social regeneration:

> It's getting back diversity, getting back choice, getting back resilience. And, my grandfather, looking through his little diary, he lived in a landscape that had a bit more resilience. They still had terrible droughts and dust storms and soil erosion events and all of that. And they went through tough times, no doubt. But they had some system resilience there, and they had choices of being able to do different things. And if they went through a bad drought and they didn't strip any grain or wheat or whatever, they could go through to the end of the next year without long-term financial impact. We can't do that now.

A fierce bushfire swept the Bethungra and Mitta Mitta districts in January 1990. Walls of flame fuelled by hot winds burned Oakley bare. Fire consumed the weathered trunks of saplings ringbarked long ago, fallen logs and branches, even humus in the topsoil. Fences, sheds, and the house were gone. 'It just looked so stark', Owen remembered.

Instead of sowing pasture seed and reintroducing stock, Owen decided to leave the burned paddocks alone. He had read about the regenerative power of Australian wildfire, and he knew something of soil seed banks and seed dormancy. When winter and spring rains came, a remarkable response unfolded. Native grasses grew so tall, Owen had to stand on the footrests of his motorbike to see ahead. Grassy woodland plants emerged that Owen had never before encountered on Oakley: yam daisies, orchids, chocolate lilies, pea bushes, and many others. He invited a botanist from Charles Sturt University at Wagga to see the extraordinary display. The scientist identified over two hundred different species.

Owen began to feel a sense of responsibility towards the rare swathe of regenerating plants, and wanted to learn how best to care for it. He knew Wiradjuri people once used fire to foster a diversity of grassy woodland species. Owen asked around, hoping to find someone with knowledge of burning practices. People directed him to an elderly man at Narrandera. Owen talked to him about Oakley and his desire to manage the land carefully. The Wiradjuri man advised Owen to return to the farm and to spend time relaxing. 'Just try and listen to your god', he said.

On reflection, Owen took the advice as a suggestion to engage closely with particular, expressive natural components and patterns active across Oakley's paddocks:

> What it meant to me was: 'Look, you bloody idiot, it's right there under your nose. Look at the land. Look at the vegetation. Observe from year to year. Look at the difference and the change in the seasons.' You know, it will tell you. The land will tell you.

Land did more than speak, Owen discovered. Over years, places worked deeply into people, shaping and holding them. During a traumatic period in the middle of the 1990s, Owen's marriage ended and Kimvale was sold. With regret, he walked away from land moulded by generations of his family, from paddocks bestowing identity.

The most difficult aspect of leaving Kimvale, Owen remembers, was seeing the effect on his children. During their childhood, hillsides and creeks gave them something profound. Active, nourishing relations between his children and the land were broken:

> I was just stunned by how much that meant to them, the landscape. And how upset they were, and just how big a part of their life it was. Yeah, it was really gut wrenching, really. Up until that they were typical teenagers. Yeah they liked the farm and liked doing stuff there, but they liked going off with their mates and they were always bitching about not being able to go to town and that. So you kind of got the impression, you know, that it doesn't really mean much to them. But I was very very wrong when it came to that. Even now, my kids say to me: 'Dad, you know, we just feel absolutely privileged to have grown up on the farm and experienced the things that we did. We just think we're so lucky. We just

had the best years we could ever have as kids there.' And that really blows me out. It just goes to show what can be the value of a landscape, outside of production. To have that effect on people. To have that connection. To provide that security.

Before the sale of Kimvale, Owen had found the experience of living in a modern farming area increasingly unpleasant. Declining terms of trade and relentlessly competitive global markets forced landholders to continually intensify farming practices and boost production. A range of environmental and social factors – the smell of agricultural chemicals, the smoke from stubble burning, the almost constant noise of diesel engines, the frantic pace of life, a declining sense of local community – contributed to his decision to relinquish Kimvale. The meanings of the farmland and paddocks had changed:

> I was really sick to my stomach of living and working in that landscape – which is not very far from here – but it had a totally different imperative to what's up here. It was a very modified landscape. I referred to it as an industrial landscape. And that may not seem too much of a surprise to many people. But, as I grew up as a kid, we never thought of it as that. Suddenly it just came to me that that's what it had become. It wasn't a natural landscape anymore. And it wasn't a family, social, or a cultural landscape. It was a straight-out industrial landscape because industrial production had become the imperative and that was driving the shape and the makeup and the type of landscape it was. And to me it was pretty ugly and difficult.

In contrast, Oakley could produce healthy sheep and cattle without heavy applications of inputs and energy, and was 'a great place to be.' Owen decided to keep Oakley, higher in the valley where his family history was deeper than on Kimvale. Remarried, he bought another farm near Yass on the southern tablelands, and moved to live there. Something more was present on the small farm he kept at Mitta Mitta, Owen knew, a significant dimension absent from modern agricultural understandings, missing even from the relatively holistic perspectives of ecological science. Walking and working on Oakley, Owen hears 'the echoes of the past.' To care for places, to keep land naturally fertile and strong, requires a sense of the past infusing the present, of time unfolding within dynamic and shifting terrains. At Oakley, ecological integrity enables such understandings:

> Here I can feel it. I can feel, you know, I can feel a small part of the short time that my white ancestors have been here. But, the majority of it I can feel just an enormous, powerful sort of a presence here, of thousands and thousands of years of human habitation and millions of years of life processes going on here – still represented, still represented in a valid way. Because there are still things happening here naturally that are totally out of my control – big things that are shaping this landscape.

Owen told me stories handed down through the generations of interaction between his family and local Wiradjuri. When Hugh and Catherine Whitaker arrived in the Mitta Mitta valley in the 1870s, Wiradjuri families lived there independently of settlers. Once, Hugh had to go away for several weeks. Soon after leaving, he realised he'd forgotten a pair of greenhide hobbles, and turned around. Arriving back at the bark hut, Hugh found Wiradjuri men threatening his wife Catherine and their children with spears, poking at them through the flimsy walls.

Owen didn't learn how the event ended, but thought violence was probably avoided. In the early years, elderly relatives told him, relationships between the Whitaker family and local Wiradjuri people were peaceful and mutually respectful. The grave of an honoured tribal chief lay on Wingana, the original Whitaker family property. Three trees with patterns carved into the trunks once surrounded the gravesite.

Wiradjuri left the district in later years as agricultural settlement intensified. Joe Whitaker, Owen's uncle, said Wiradjuri used to return to Mitta Mitta occasionally to visit the grave. Joe told a story of his uncles, as boys, deciding to open the chief's grave and unearth weapons the family understood were buried alongside the dead warrior. Digging, they saw ghostly faces emerge from nearby trees. The boys fled. Decades later, only one of the three carved trees remained, dead and decaying. Joe Whitaker carefully removed the decorated section of heartwood and gave it to the Gundagai museum for safekeeping, along with spears and woomeras collected at Mitta Mitta and a photograph of the dead tree. His daughter Elizabeth told Owen the family always respected local Wiradjuri. Joe Whitaker never allowed his children to go near the warrior's gravesite.

Beside the creek on Oakley, Owen finds stone flakes and blades. Some of the tools look fresh and clean, he explained, as if made only a year or two ago. Others are ancient, weathered. Like narratives of encounters with Wiradjuri passed down through family generations, Owen considers the stones significant partly because they represent a tenuous bond with alternative ways of seeing and understanding that are relevant today. Above the creek running with spring water, on the hillside where we talked, Wiradjuri saw and carefully tended dynamic and expressive patterns of life.

Regeneration of rural places requires perceptions of people as constructive members of biological communities. People may care for local ecologies – enable natural diversity, connectivity, strength and productivity – and be nourished in return. Responsive, careful relationships require intimate knowledge of places. Western notions of a firm divide between people and the rest of nature impede dialogue and empathy. Feelings of belonging and connection to the vibrant life of places are denied.

Likewise, understandings of the present as severed from the consequences of historical events block acceptance of responsibility for social and ecological wellbeing. Always active in the present, the past infuses all places and lives and shapes our complex realities.

At the Temora Rural Museum, northwest of Mitta Mitta,

stone axe heads collected in local paddocks are displayed in a glass cabinet. Alongside historic and more contemporary farm equipment, the stone tools help construct simplistic narratives of linear progress from primitive Aboriginal pasts to sophisticated settler futures. In contrast, the stone flakes and blades found at Oakley, as described by Owen, offer possibilities to dismantle problematic dichotomies of western belief: past versus future, primitive versus European, wild versus civilised, body versus mind, nature versus culture, land versus people, rural versus urban. The regeneration and enduring health of rural places and agricultural regions requires fresh narratives and perceptions to undermine and replace inappropriate and destructive traditions of thought and culture.

Geographic and cultural divides between rural and urban domains keep the ecological wounds of farming regions open and festering. Farmers need support to counter and subvert the powerful, corrosive processes of economic fundamentalism, industrial production and globalised trade. The formidable and promising task of regenerating the agricultural heartlands of Australia cannot be borne by them alone. There is potential for city residents to turn towards adjacent rural hinterlands. Seeking dialogue and close ties with nearby rural places and people, urban dwellers may sense personal and shared embeddedness within the histories and natural systems of their local regions. A profound style of caring for Australian farmland may develop as people take responsibility for the wellbeing of places surrounding and nourishing them.

Acceptance of ethical obligations to care for nearby farmlands promises more than ecological and social regeneration. Rejection of industrial and colonial mindsets enables emotional and sensual engagement with natural particularities and patterns. As Owen Whitaker and his children learned at Mitta Mitta, lively places offer people rich gifts: a deeper experience of our humanity, and land to cherish and know intimately as home.

notes

Introduction

1 'Cootamundra and district, a *Land* special feature', *The Land*, 18 June 1970.
2 Department of Land & Water Conservation, *Murrumbidgee catchment stressed rivers assessment report*, August 1999, p. 57.
3 Deborah Rose, Diana James & Christine Watson, *Indigenous kinship with the natural world in New South Wales*, NSW National Parks & Wildlife Service, Hurstville, 2003.
4 AW Howitt, *The native tribes of south-east Australia*, Aboriginal Studies Press, Canberra, 1996 [1904], p. 56.
5 Mary Gilmore, *Old days: old ways – a book of recollections*, Angus & Robertson, Sydney, 1963 [1934], p. 145.
6 'British millers' requirements in wheat', *Agricultural Gazette of New South Wales*, vol. 9, July to December 1898, p. 750.
7 Heather Goodall, '"Fixing" the past: modernity, tradition and memory in rural Australia', *UTS Review*, vol. 6, no. 1, May 2000, p. 22.
8 James C Scott, *Seeing like a state: how certain schemes to improve the human condition have failed*, Yale University Press, New Haven, 1998, p. 275.
9 Fritjof Capra, *The web of life: a new synthesis of mind and matter*, Flamingo, London, 1997, pp. 184–5.
10 Jill Ker Conway, Kenneth Keniston & Leo Marx, *Earth, air, fire, water: humanistic studies of the environment*, University of Massachusetts Press, Amherst, 1999, p. 3.
11 Wendell Berry, *Life is a miracle: an essay against modern superstition*, Counterpoint, Washington, 2001, p. 88.
12 Tom Griffiths, 'The humanities and an environmentally sustainable Australia', address to Department of Education, Science and Training conference, 'The contribution of the social sciences and humanities to national research priorities', Canberra, 28 March 2003.
13 Donald Worster, *The wealth of nature: environmental history and the ecological imagination*, Oxford University Press, New York, 1993, p. 178.
14 James Gormly, 'Early Tarcutta history', Gormly family records, Charles Sturt University Regional Archives, Wagga Wagga.
15 Ernest Fletcher, *Tarcutta centenary 1936*, St Mark's Church, Tarcutta, 1936.

16 Jane Franklin, 1839 diary, National Library of Australia, manuscript collection, MS 114, entry for 25 April 1839.
17 Justin Nancarrow, Robert Cawley & Tim Smith, *The ecological health of Tarcutta Creek: a study of the physical and biological integrity of Tarcutta Creek*, Department of Land & Water Conservation, Murrumbidgee Region, 2001.
18 Deborah Rose, 'Rupture and the ethics of care in colonized space', in Tim Bonyhady & Tom Griffiths, *Prehistory to politics: John Mulvaney, the humanities and the public intellectual*, Melbourne University Press, Carlton, 1996.

Mastery

1 Sarah Musgrave, *The wayback*, Bland District Historical Society & Young Historical Society, West Wyalong, 1979 [1926], pp. 2–3.
2 Heather Goodall, *Invasion to embassy: land in Aboriginal politics in New South Wales, 1770–1972*, Allen & Unwin, St Leonards, 1996, p. 63.
3 Peter Read, *A hundred years war: the Wiradjuri people and the state*, privately published, Canberra, 1994 [1988], p. 10.
4 Charles Sturt, *Proceedings of an expedition into the interior of New Holland, 1829 and 1830*, Sullivan's Cove, Adelaide, 1989, p. 18.
5 Stephen Roberts, *The squatting age in Australia 1835–1847*, Melbourne University Press, Melbourne, 1964 [1935], p. 81.
6 Henry Bingham to Edward Deas Thompson, 26 October 1839, State Records NSW, 4/2439.1.
7 Henry Bingham to Edward Deas Thompson, 21 July 1839, State Records NSW, 4/2438.2.
8 Colonial Secretary's Office, *New South Wales Government Gazette*, Sydney, 21 August 1839.
9 James Gormly, 'Early days in and near Wagga Wagga', Gormly family records, Charles Sturt University Regional Archives, Wagga Wagga.
10 Henry Cosby, evidence given before the Committee on Police and Gaols, 'Report of the Committee on Police and Gaols', *Votes and proceedings*, vol. 2, NSW Legislative Council, Sydney, 1839, p. 69.
11 Gilmore, *Old days: old ways*, pp. 70–1.
12 Gilmore, *Old days: old ways*, p. 71.
13 Thomas Livingstone Mitchell, *Journal of an expedition into the interior of tropical Australia in search of a route from Sydney to the Gulf of Carpentaria*, Friends of the State Library of South Australia, Adelaide, 1999 [1848], p. 414.
14 Jennifer Baldry, *Swords scythes and shears: a history of the sawyers of Bethungra New South Wales and related families*, privately published, Wallendbeen, 1989, p. 61.
15 Colonial Secretary's Office, 'Claims to leases of crown land', *Supplement to the New South Wales Government Gazette*, 27 September 1848, p. 1318.
16 'Amended descriptions of the Tymora and Rock runs', 15 March 1869, copy held by Jim Main, Cootamundra.

17 Henry Cosby to Edward Deas Thompson, May 1839, State Records NSW, 4/2438.2.
18 Bill Gammage, *Narrandera Shire*, Bill Gammage for the Narrandera Shire Council, Narrandera, 1986, pp. 17–20; 'Ngarrang' denotes a particular lizard species, either a jew lizard (*Pogona barbata*) or the eastern water dragon (*Physignathus lesueurii*). Ngarrangdhuray gave sanctuary to lizards at Narrandera, in the same way Gudhamangdhuray clanspeople nurtured turtle populations at Cootamundra.
19 Gammage, *Narrandera Shire*, pp. 32–4.
20 Henry Cosby to Edward Deas Thompson, 12 May 1839, State Records NSW, 4/2438.2.
21 James Gormly, *Exploration and settlement in Australia*, Ford, Sydney, 1921, p. 118.
22 Gammage, *Narrandera Shire*, pp. 35–8.
23 James Baylis, 'The Murrumbidgee and Wagga Wagga', *Journal and Proceedings*, Royal Australian Historical Society, vol. 13, 1927, p. 256.
24 James Devlin Sr, Stan Devlin & Elsie Devlin, *Reminiscences*, privately published volume in the Wagga Wagga City Library, no date.
25 Devlin, *Reminiscences*.
26 Keith Swan, *A history of Wagga Wagga*, City of Wagga Wagga, Wagga Wagga, 1970, p. 13.
27 Wagga Wagga and District Historical Society newsletter, no. 58, December 1967, p. 2.
28 'Sebastopol in seventy: Mr. H.M. Kavanagh tells an interesting story', in Watson A Steele, *Temora's jubilee souvenir (illustrated): containing history of Temora*, JA Bradley, Temora, 1931.
29 Edgar Beckham, 'Report upon the condition, etc., of the Aborigines inhabiting the Lachlan District during the year 1844', in Frederick Watson (ed.), *Historical records of Australia*, series 1, vol. XXIV, Library Committee of the Commonwealth Parliament, 1925, pp. 267–8.
30 George Bennett, *Wanderings in New South Wales*, vol. 1, Richard Bentley, London, 1834, p. 305.
31 Mary Gilmore, 'The hunter of the black', in *The passionate heart and other poems*, Angus & Robertson, 1969 [1948], pp. 66–8.
32 GL Buxton, *The Riverina 1861–1891: an Australian regional study*, Melbourne University Press, Carlton, 1967, p. 148.
33 Deborah Rose, *Nourishing terrains: Australian Aboriginal views of landscape and wilderness*, Australian Heritage Commission, Canberra, 1996, p. 49.
34 Gilmore, *Old days: old ways*, p. 118.
35 Gilmore, *The passionate heart*, p. 307.
36 Gilmore, *Old days: old ways*, p. 168.
37 Gilmore, *Old days: old ways*, pp. 153–4.
38 Gilmore, *Old days: old ways*, pp. 154–5.
39 Mary Gilmore, 1940 diary, National Library of Australia, manuscript collection, MS 614, entry for 12 February 1940.
40 Gilmore, *The passionate heart*, p. 315.
41 Gilmore, *Old days: old ways*, pp. 117–18.
42 Gilmore, *Old days: old ways*, p. 140.

43 Gilmore, *Old days: old ways*, p. 119.
44 Gilmore, *Old days: old ways*, p. 152.
45 Gilmore, *Old days: old ways*, pp. 118–20.
46 Gilmore, *The passionate heart*, pp. 317–18.
47 Colonial Secretary's Office, 'Claims to leases of crown land beyond the settled districts. Lachlan District', *Supplement to the New South Wales Government Gazette*, Wednesday, 27 September 1848, p. 1315.
48 Richard Littlejohn, *An Australian pioneer: Alexander Mackay 1815–1890*, privately published, Wallendbeen, 1992, p. 20.
49 'The southern districts', *Sydney Morning Herald*, 30 January 1844; and 'Aborigines', *Gormly Index*, Charles Sturt University Regional Archives, Wagga Wagga.
50 Beckham, 'Report upon the condition of the Aborigines', pp. 267–8.
51 Musgrave, *The wayback*, p.17.
52 Littlejohn, *An Australian pioneer*, p.20.
53 Frank Clune, *Last of the Australian explorers: the story of Donald Mackay*, Angus & Robertson, Sydney, 1942, p.17.
54 Kenneth Mackay, 'Pioneers of 1837–60', *Sydney Mail*, 3 January 1923.
55 Draft of a letter from Alexander Mackay, 4 March 1849, Baldry family papers.
56 Gormly, *Exploration and settlement in Australia*, p.200.
57 'A tour through the pastoral district of the Bland', *Cootamundra Herald*, 11 January 1879.
58 Veronica G McNamara, *Beyond the early maps*, privately published, Orange, 1974, p.177.
59 McNamara, *Beyond the early maps*, p.94.
60 Eric Rolls, *A million wild acres*, Penguin Books, Ringwood, 1984, p.164.
61 James Günther, 'Grammar and vocabulary of the Aboriginal dialect called the Wirradhuri', appendix to Lancelot Edward Threlkeld, *An Australian language as spoken by the Awakabal*, Charles Potter, Government Printer, 1892.
62 Mitchell, *Journal of an expedition into the interior of tropical Australia*, pp. 412–13.
63 Royal Commission of Inquiry on Forestry, *Minutes of proceedings, minutes of evidence, and appendix*, Part II, William Applegate Gullick, Government Printer, Sydney, 1908, pp. 535–6.
64 Legislative Assembly of New South Wales, 'Clearing pine and scrub from leased lands', *Votes and proceedings*, vol. 3, Legislative Assembly of New South Wales, Sydney, 1881, pp. 252–66.
65 Rob Webster, *The first fifty years of Temora*, JA Bradley, Temora, 1950, p.13.
66 George L Sutton, *Wheat-growing in New South Wales*, William Applegate Gullick, Sydney, 1908, pp. 12–13; WS Ramson (ed.), *The Australian national dictionary*, Oxford University Press, Melbourne, 1988, p. 646.
67 'The value of ringbarking', *The Pastoral Review*, 15 July 1939, p. 767.
68 William Farrer, *Grass and sheep-farming*, William Maddock, Sydney, 1873.
69 Sutton, *Wheat-growing in New South Wales*, p. 14.
70 DJ Halliday, *The wheat-growers of south-western New South Wales to 1914*, MA thesis, Department of History, University of Sydney, 1962, p. 111.

71 'New tree-extractor', *Cootamundra Herald*, 2 August 1884.
72 Mary Gilmore, *Hound of the road*, Angus & Robertson, Sydney, 1922, pp. 44–6.
73 Brett J Stubbs, 'Land improvement or institutionalised destruction? The ringbarking controversy 1879–1884, and the emergence of a conservation ethic in New South Wales', *Environment and History*, vol. 4, no. 2, June 1998, p. 149.
74 'Our studmasters. Mr. Frank Cowley, Bethungra Park', *Sydney Mail*, 13 December 1884.
75 Pastoralists' Review Pty Ltd, *The pastoral homes of Australia*, Sydney, 1910.
76 Colin M Donald, 'Innovation in Australian agriculture', in DB Williams (ed.), *Agriculture in the Australian economy*, Sydney University Press, Sydney, 1982, p. 59.
77 *Productive solutions to dryland salinity*, Grains Research and Development Corporation, Kingston, 2001.
78 'Research addresses waterlogging woes', *Rural News*, 12 October 2001.
79 'Old hand's tale pleases writer', *Forbes Advocate*, 19 July 1955.
80 Mackay, 'Pioneers of 1837–60'.
81 Ethel West, 'Stockinbingal: full belly or marshland?', *Cootamundra Herald*, 6 July 1973.
82 James Larmer, 1848 field book, State Records NSW, 2/8070.1
83 'A tour through the pastoral district of the Bland'.
84 Gilmore, *The passionate heart*, p. 319.
85 'Early memories of Mrs. S.K. Hawkins, of Curraburrama', in *Pink hats on gentle ladies*, family history notes compiled in 1974 by Vida Clift and transcribed from the *Young Chronicle*, 4 November 1932.
86 Colin M Donald, 'The progress of Australian agriculture and the role of pastures in environmental change', *Australian Journal of Science*, vol. 27, no. 7, January 1965, p. 187.
87 Bruce R Davidson, 'The history of agriculture in temperate Australia', *Journal of the Australian Institute of Agricultural Science*, vol. 51, no. 3, 1985, p. 165.
88 Robert D Watt, *The romance of the Australian land industries*, Angus & Robertson, Sydney, 1955, pp. 214–15.
89 Robert Jackson Noble, Bright Sparcs biographical entry, accessed February 2003, <http://www.asap.unimelb.edu.au/bsparcs/biogs/P000675b.htm>.
90 Robert Jackson Noble, *Agriculture and the United Nations*, 1959 Farrer Memorial Oration, VCN Blight, Government Printer, Sydney, 1960, p. 8.
91 NSW Agriculture, 'Brief history', accessed March 2003, <http://www.agric.nsw.gov.au/reader/35>.
92 June Sutherland, *From farm boys to PhDs: agricultural education at Wagga Wagga*, Charles Sturt University, Wagga Wagga, 1996, pp. 2–3.
93 NSW Agriculture, 'Temora Agricultural Research and Advisory Station', accessed April 2002, <http://www.agric.nsw.gov.au/reader/10855>.
94 Neil Barr & John Cary, *Greening a brown land: the Australian search for sustainable land use*, Macmillan, Melbourne, 1992, p. 34.
95 Greg Whitwell, *The Australian wheat industry: 1939–1989*, Macmillan Australia, Melbourne, 1991, p. 67.

96 Eric J Underwood, 'The role of science in the development of the animal industries of Australia', AS Nivison Memorial Address, delivered at the University of New England, Armidale, on 19 October 1967, published version held by the National Library of Australia; and Donald, 'The progress of Australian agriculture'.
97 David F Smith, *Natural gain in the grazing lands of southern Australia*, UNSW Press, Sydney, 2000, pp. 201–5.
98 David F Smith, 'Caring for the land – for ever and ever', paper delivered at the Murrumbidgee Landcare Association Forum, Cootamundra, 2000.
99 National Farmers' Federation, *Australian agriculture*, Melbourne, 1985, p. 175.
100 David Tacey, 'Dissolving into landscape: psyche, earth & sacrifice', *Island*, no. 56, spring 1993, p. 48.
101 Underwood, 'The role of science', pp. 10, 12.
102 Underwood, 'The role of science', p. 3.
103 Robert Pogue Harrison, *Forests: the shadow of civilisation*, University of Chicago Press, Chicago, 1992, pp. 107–8.
104 'Yanco Agricultural High School: some early history', accessed April 2003, <http://www.yahs.nsw.edu.au/history.htm>.
105 'Development of agricultural education in N.S.W.', *Sydney Mail*, 17 January 1923.
106 Otto Herzberg Frankel, Bright Sparcs biographical entry, accessed October 2003, <http://www.asap.unimelb.edu.au/bsparcs/biogs/P000422b.htm>; and Otto Herzberg Frankel, 'The social responsibility of agricultural science', Farrer Memorial Oration, 1962, *Australian Journal of Science*, vol. 25, no. 7, January 1963, p. 301.
107 Smith, *Natural gain in the grazing lands of southern Australia*, pp. 163, 174.
108 Eric J Underwood, papers, National Library of Australia, manuscript collection, MS 7082.
109 Underwood, 'The role of science', p. 14.
110 Geoffrey Lawrence & Ian Grey, 'The myths of modern agriculture: Australian rural production in the 21st century', in Bill Pritchard & Phil McManus (eds), *Land of discontent: the dynamics of change in rural and regional Australia*, UNSW Press, Sydney, 2000, p. 38.
111 Vaughan Higgins & Stewart Lockie, 'Getting big and getting out: government policy, self-reliance and farm adjustment', in Stewart Lockie & Lisa Bourke (eds), *Rurality bites: the social and environmental transformation of rural Australia*, Pluto Press, Sydney, 2001, pp. 178–90.
112 'Agricultural societies and farmers' clubs', *The Garden and the Field: A Journal of General Industries*, vol. 4, no. 39, 1 August 1878.
113 Wilcox Mofflin Limited, 'The quickest, surest way to solve weed problems', *Agricultural Gazette of New South Wales*, May 1955, opposite p. 268.
114 'Rust, grubs, aphids, mice', *The Land*, 7 October 2004.
115 Henty Centenary Committee, *From early beginnings: Henty NSW, home of the header*, Henty, 1986, p. 276.
116 'Farm machinery display grows', *Rural News*, 7 September 2001.

117 'Henty Machinery Field Days preview', *The Land*, 11 September 2003.
118 'Welcome to Henty Machinery Field Days Online', accessed November 2003, <http://www.hmfd.com.au>.
119 Australian Bureau of Statistics, 'Agriculture: financial statistics of farm businesses', *Year Book Australia 2003*, Canberra, 2003.
120 Henty Centenary Committee, *From early beginnings*, pp. 59, 172.

Elsewhere

1 Gilmore, *Old days: old ways*, pp. 130–33.
2 Gammage, *Narrandera Shire*, p. 18.
3 Musgrave, *The wayback*, pp. 8–9.
4 Letter from Wallendbeen, 31 December 1850, Baldry family papers.
5 Gammage, *Narrandera Shire*, p. 214.
6 Gormly, *Exploration and settlement in Australia*, pp. 164–6.
7 Gormly, 'Early days in this district', Gormly family records, Charles Sturt University Regional Archives, Wagga Wagga.
8 Gammage, *Narrandera Shire*, p. 53.
9 Sherry Morris, *Wagga Wagga: a history*, Council of Wagga Wagga, Wagga Wagga, 1999, pp. 76–7.
10 Eric Irvin (ed.), *Letters from the river*, Wagga Wagga, 1959.
11 Gammage, *Narrandera Shire*, p. 35.
12 Richard Littlejohn, *Early Murrumburrah historical notes*, Harden–Murrumburrah Historical Society, Harden–Murrumburrah, no date, pp. 5–6.
13 *Cootamundra Herald*, 20 March 1877.
14 'Opening of railway', *Cootamundra Herald*, 6 November 1877.
15 Selected and adapted from a report of the banquet, *Cootamundra Herald*, 6 November 1877.
16 'Opening of the railway to Junee', *Cootamundra Herald*, 9 July 1878.
17 Morris, *Wagga Wagga*, p. 77.
18 WJ Garland, *The history of Wagga Wagga*, Riverina College of Advanced Education, Wagga Wagga, 1984 [1913], p. 6.
19 'A new railway line', *Town and Country Journal*, 19 August 1893.
20 *Cootamundra Herald*, 4 December 1877.
21 Halliday, *The wheat-growers of south-western New South Wales*, pp. 107–8.
22 'Trial of agricultural machinery', *Cootamundra Herald*, 4 December 1877.
23 *Town and Country Journal*, 2 January 1907.
24 'Progress of wheat-growing in New South Wales', *Town and Country Journal*, 30 June 1909.
25 'Through progressive districts – Cootamundra to Barellan', *Town and Country Journal*, 30 June 1909.
26 Webster, *The first fifty years of Temora*, p. 80.
27 'Australia's finest range of health foods', *Australia to-day*, United Commercial Travellers' Association of Australasia, Melbourne, 1964, p. 91.
28 Freya Matthews, 'Letting the world do the doing', *Australian Humanities Review*, Issue 33, August–October 2004.
29 John Stuart Mill, quoted in Edward W Said, *Culture & imperialism*, Vintage,

London, 1994, p. 108.
30 Geoffrey Bolton, *Spoils and spoilers: Australians make their environment 1788–1980*, Allen & Unwin, Sydney, 1981, p. 21.
31 William Creser, *The breaking of the drought*, Massey-Harris, London, 1904, copy held by the National Library of Australia.
32 Frank Sweeny, 'Early settlers at Cullinga', as dictated to his niece Mrs Dixon, Harden–Murrumburrah Historical Society bulletins 1974–1975.
33 Patricia Caskie, *Cootamundra 1901–1924: past imperfect*, Annwel Enterprises, Cootamundra, 2000, p. 373.
34 *Wagga Wagga Advertiser*, 3 September 1924.
35 Address to Sir Dudley de Chair, Mitchell Library, reference number *D184.
36 'Coota landmark under construction', *Cootamundra Herald*, 3 July 1981.
37 'Wheat exports to rise despite lower yields', *Sydney Morning Herald*, 13 December 2004.
38 'Hallam opens $2m wheat bins', *Cootamundra Herald*, 29 September 1982.
39 Greg Whitwell, *A shared harvest: the Australian wheat industry, 1939–1989*, Macmillan, Melbourne, 1991, p. 69.
40 Bill Pritchard, 'Negotiating the two-edged sword of agricultural trade liberalisation: trade policy and its protectionist discontents', in Pritchard & McManus, *Land of discontent*, pp. 90–104.
41 Vandana Shiva, 'The violence of globalisation', *The Hindu*, 25 March 2001.
42 Paul B Thompson, *The spirit of the soil: agriculture and environmental ethics*, Routledge, London, 1995, p. 48.
43 Geoffrey Lawrence, 'Who controls our future?' *Proceedings of the 9th Australian Agronomy Conference*, Wagga Wagga, 1998.
44 Aldo Leopold, *A Sand County almanac and sketches here and there*, Oxford University Press, New York, 1987 [1949], p. 214.
45 Australian Taxation Office, *Income Tax Assessment Act 1936*, Section 75A, 'Deduction of certain expenditure on land used for primary production', accessed February 2003, <http://law.ato.gov.au/atolaw/view.htm>.
46 *Murrumbidgee catchment stressed rivers assessment report*, p. 57.
47 Commonwealth of Australia, *Managing natural resources in rural Australia for a sustainable future: a discussion paper for developing a national policy*, December 1999, accessed July 2003, <http://www.napswq.gov.au/publications/nrm-discussion.html>.
48 'Increasing farmer profitability the aim of Patterson Lott Associates', *Southern Weekly Magazine*, 14 April 2003.
49 Philip Knopke, Ben Furmage, Peter Walters & Anthony King, 'Agriculture in the Australian economy', p. 28, in National Farmers' Federation, *Australian agriculture*, pp. 17–33.
50 Susan Butler (ed.), *The Macquarie dictionary of new words*, The Macquarie Library, Sydney, 1990, p. 353.
51 JA Simpson & ESC Weiner, *The Oxford English dictionary*, vol. 17, Oxford University Press, 1989, pp. 326–7.
52 Ian Macdonald, 'Agriculture on the move', *NSW Agriculture Today*, 28 August 2003.
53 'Sustaining the land', accessed June 2003, <http://www.nswfarmers.org.au/conservation/casestudies>.

54 Smith, 'Caring for the land'.
55 Warren Truss, address to the International Landcare 2000 Conference, 3 March 2000, accessed November 2003, <http://www.affa.gov.au/ministers/truss/speeches/ilcspeech.html>.
56 Australian Bureau of Agricultural and Resource Economics (ABARE), 'Trend toward larger farms likely to continue', media release, 11 July 2002, accessed November 2003, <http://www.abare.gov.au/pages/media/2002/11July1%20.htm>.
57 Wes Jackson, *Altars of unhewn stone: science and the earth*, North Point Press, New York, 1987, pp. 15–16.
58 Wendell Berry, *What are people for?* North Point Press, San Francisco, 1990, pp. 206–7.
59 Wendell Berry, *Life is a miracle*, p. 134.

Progress

1 Evelyn Sturt, letter to Charles La Trobe, 20 October 1853, in Thomas Francis Bride (ed.), *Letters from Victorian pioneers*, Lloyd O'Neil, Melbourne, 1983, p. 364.
2 Gilmore, *Old days: old ways*, p. 146.
3 Gilmore, *Old days: old ways*, pp. 117–18.
4 'Through progressive districts – Cootamundra to Barellan'.
5 'Death of Mr. Alexander Mackay', *Cootamundra Herald*, 8 February 1890.
6 Kenneth Mackay, *Songs of a sunlit land*, Angus & Robertson, Sydney, 1908, pp. 14–16.
7 *The pastoral homes of Australia*.
8 Michael E Robinson, *The New South Wales wheat frontier 1851 to 1911*, Research School of Pacific Studies, Canberra, 1976, p. 203.
9 Peter Read, 'Fathers and sons: a study of five men of 1900', *Aboriginal History*, vol. 4, 1980, p. 99.
10 'The Warangesda Mission', *Cootamundra Herald*, 4 February 1882.
11 'How the cannibals presented me with a wife', *Cootamundra Herald*, 21 June 1890.
12 'Land for selection', *Cootamundra Herald*, 25 June 1890.
13 'Cootamundra: a great record of progress', *Cootamundra Herald*, 25 January 1896.
14 'Progress committee', *Cootamundra Herald*, 8 March 1890.
15 *Town and Country Journal*, 30 June 1909.
16 'Cootamundra. A quarter-century's growth', *Sydney Mail*, 22 December 1900.
17 Michael Davitt, *Life and progress in Australasia*, Methuen, London, 1898, p. 175.
18 Tom Griffiths, *Beechworth: an Australian country town and its past*, Greenhouse Publications, Richmond, 1987, p. 1.
19 B Boddington (ed.), *Official souvenir of Cootamundra centenary 1861–1961*, pamphlet, Cootamundra, 1961.
20 A petition to the Honourable PH Morton, MLA, Minister for Local

Government and Minister for Highways on the Town Hall, Cootamundra, 1969, held by Cootamundra Library.
21 'Decision to demolish hall facade', *Cootamundra Herald*, 26 March 1969.
22 '"Taking us into 21st century," says mayor', *Cootamundra Herald*, 11 June 1971.
23 'New buildings also serve a town's history', *Cootamundra Herald*, 11 June 1971.
24 Dipesh Chakrabarty, *Provincializing Europe: postcolonial thought and historical difference*, Princeton University Press, Princeton, 2000, p. 244.
25 *Cootamundra: an invitation to discover*, Cootamundra Development Corporation, 2004.
26 Cootamundra Shire Council, 'General information', accessed August 2003, <http://cootamundra.local-e.nsw.gov.au/about/01/1056326289_19124.html>.
27 Kevin Gilbert, *The cherry pickers*, Burrambinga Books, Canberra, 1988.
28 Steele, *Temora's jubilee souvenir*.
29 '*Temora Independent* centenary celebrations supplement', *Temora Independent*, Temora, 1980.
30 Colin M Donald, 'The progress of Australian agriculture and the role of pastures in environmental change', *Australian Journal of Science*, vol. 27, no. 7, January 1965.
31 Vandana Shiva, *The violence of the green revolution: Third World agriculture, ecology and politics*, Zen Books, London, 1991, p. 24.
32 'Non-start to water debate', *The Land*, 31 October 2002.
33 Smith, *Natural gain in the grazing lands of southern Australia*, p. 11.
34 Peter Read, *a history of the Wiradjuri people of New South Wales 1883–1969*, PhD thesis, Australian National University, 1983, pp. 231–3.
35 David Abram, *The spell of the sensuous: perception and language in a more-than-human world*, Vintage, New York, 1997, pp. 216–17.
36 Abram, *The spell of the sensuous*, p. 272.
37 Edward W Said, *Culture and imperialism*, Vintage, London, 1994, p. 159.
38 'Water supply', *Cootamundra Herald*, 14 November 1885.
39 'Important discovery', *Cootamundra Herald*, 14 December 1881.
40 'Water supply', *Cootamundra Herald*, 8 February 1882.
41 'River water is officially turned on', *Cootamundra Herald*, 24 February 1933.
42 Gormly, 'Early days in this district'.
43 'Huge crowd expected at park opening', *Cootamundra Herald*, 23 July 1971.
44 Peter Rimas Kabaila, *Wiradjuri places: the Murrumbidgee River basin with a section on Ngunawal country*, 2nd edn, Black Mountain Projects, Canberra, 1998, p. 74; and Read, *A hundred years war*, pp. 37–8.
45 Read, *A hundred years war*, pp. 35–7.
46 Vince Bulger, quoted in Kabaila, *Wiradjuri places*, pp. 83–6.
47 Gormly, 'Early days in this district'.
48 Arthur McGuiness, taped conversation with Janet Mathews, Brungle, c.1965, tape held by Bob Glanville.
49 Read, *A hundred years war*, p. 64.
50 Margaret Tucker, *If everyone cared: autobiography of Margaret Tucker MBE*, Grosvenor, London, 1983, p. 82.

51 Kabaila, *Wiradjuri places*, p. 59.
52 Office of the Director of Equal Opportunity in Public Employment, *Yarnin' up: Aboriginal people's careers in the NSW public sector*, 2001.
53 Leopold, *A Sand County almanac*, p. 223.
54 Val Plumwood, *Environmental culture: the ecological crisis of reason*, Routledge, London, 2002, p. 43.

Division

1 Hugh Brody, *The other side of Eden: hunter-gatherers, farmers and the shaping of the world*, Faber & Faber, London, 2001, pp. 306–7.
2 Freya Mathews, 'Ceres: singing up the city', *PAN*, no. 1, 2000, p. 9.
3 Avcare, 'Crop protection', accessed October 2003, <http://www.avcare.org.au/default.asp?V_DOC_ID=769>.
4 Rachel Carson, *Silent spring*, Penguin, London, 1991 [1965], p. 27.
5 Adrian Piccoli, *Drought: a survival guide*, Murrumbidgee Electorate Office, Griffith, 2002.
6 Dedee Woodside, on 'The great water debate', *Bush Telegraph*, ABC Radio National, 13 November 2002.
7 Kay Hull, in Daniel Connell (ed.), *Uncharted waters*, Murray–Darling Basin Commission, Canberra, 2002, p. 17.
8 'Nobel Peace Prize winner backs revolution', accessed June 2003, <http://www.grdc.com.au/whats_on/mr/borlaug_april2003.htm>.
9 Shiva, *The violence of the green revolution*, p. 248.
10 John Williams, 'Towards sustainable land management', *Food for healthy people and a healthy planet*, internet conference, September 2001, accessed February 2002, <http://www.natsoc.org.au>.
11 Shiva, *The violence of the green revolution*, p. 256.
12 Donald Worster, *Dust bowl: the southern plains in the 1930s*, Oxford University Press, New York, 1979, p. 66.
13 Plumwood, *Environmental culture*, p. 104.
14 Patricia Seed, *Ceremonies of possession in Europe's conquest of the New World, 1492–1640*, Cambridge University Press, Cambridge, 1995, pp. 16–29.
15 Richard Aitken & Michael Looker (eds), *The Oxford companion to Australian gardens*, Oxford University Press, South Melbourne, 2002, p. vii.
16 Paul Fox, 'Homestead gardens', in Aitken & Looker, *The Oxford companion to Australian gardens*, pp. 308–9.
17 Neil Robertson in Neil Conning & Associates (eds), *Australia's open garden scheme guidebook: 2000/2001*, ABC Books, 2001.
18 Fox, 'Homestead gardens', p. 309.
19 Val Plumwood, 'Decolonising relationships with nature', *PAN*, no. 2, 2002, p. 14.
20 Lynne Landy, 'Launch of the 2002/2003 season of Australia's open garden scheme', accessed June 2003, <http://www.governor.vic.gov.au/4A25671B00145208/WebMrsLynneSpeeches/F58AB590E8738CDECA256CD7000A4186?OpenDocument>.
21 Seed, *Ceremonies of possession*, p. 28.

22 Judith Wright, *Born of the conquerors: selected essays by Judith Wright*, Aboriginal Studies Press, Canberra, 1991, pp. 18–19.
23 Plumwood, 'Decolonising relationships with nature', pp. 10–11.
24 Watt, *The romance of the Australian land industries*, p. 202.
25 Richard Manning, 'The oil we eat: following the food chain back to Iraq', *Harper's Magazine*, February 2004.
26 Victoria Post Ranney (ed.), *An open land: photographs of the midwest, 1852–1982*, Open Lands Project, Chicago, 1983.
27 Plumwood, *Environmental culture*, p. 103.
28 'On the land', *Garden design*, Australian House & Garden, Sydney, 1992.
29 HJ Wedge, *Wiradjuri spirit man*, Craftsman House, Sydney, 1996, p. 9.
30 Rose, *Nourishing terrains*.
31 Wedge, *Wiradjuri spirit man*, p. 80.
32 Leopold, *A Sand County almanac*, pp. 203–4.
33 *Cootamundra Herald*, 20 August 1878.
34 'Some southern towns. Cootamundra described', *Freeman's Journal*, 16 June 1900.
35 Simon Ryan, *The cartographic eye: how explorers saw Australia*, Cambridge University Press, Melbourne, 1996, p. 62.
36 Jonathan Bate, *The song of the earth*, Picador, London, 2000, p. 132.
37 Richard Flanagan, *Gould's book of fish: a novel in twelve fish*, Picador, Sydney, 2001, p. 93.
38 Ryan, *The cartographic eye*, p. 57.
39 Sturt, *Proceedings of an expedition into the interior of New Holland*, pp. 19–20.
40 Sturt, *Proceedings of an expedition into the interior of New Holland*, p. 22.
41 Ryan, *The cartographic eye*, p. 80.
42 National Farmers' Federation, *Australian agriculture*, p. 41.
43 Scott, *Seeing like a state*, p. 281.
44 Stephen Muecke, 'A landscape of variability', in Martin Thomas (ed.), *Uncertain ground: essays between art and nature*, Art Gallery of New South Wales, Sydney, 1999, p. 48.
45 'Experienced, field proven performers', *The Furrow*, John Deere Ltd, Neutral Bay Junction, no. 3, 2002.
46 Steele, *Temora's jubilee souvenir*.
47 James Gormly, 'Early days in this district: the back stations', Gormly family records, Charles Sturt University Regional Archives, Wagga Wagga.
48 Mary Gilmore, *More recollections*, Angus & Robertson, Sydney, 1935, p. 15.
49 Marjorie Barnard Eldershaw, *My Australia*, Jarrolds, London, 1939, pp. 142–3.
50 John Rickard, *Australia: a cultural history*, Longman, New York, 1992, p. 69.
51 JJ Keenan, *The inaugural celebrations of the Commonwealth of Australia*, NSW Government, Sydney, 1904, pp. 29–37.
52 Russel Ward, *The Australian legend*, Oxford University Press, Melbourne, 1958, pp. 1–2.
53 Don Aitken, '"Countrymindedness" – the spread of an idea', *Australian*

Cultural History, no. 4, 1985, pp. 34–40.
54 Lockie & Bourke, *Rurality bites*, pp. 9–10.
55 Geoffrey Blainey, *2001 Boyer Lectures*, 'Lecture 3: The great divide', accessed January 2004, <http://www.abc.net.au/rn/boyers/stories/s420929.htm>.
56 Julian Cribb, *The forgotten country*, Australian Farm Publications, Melbourne, 1982, p. 7.
57 Lawrence & Gray 'The myths of modern agriculture', pp. 36–7.
58 NSW Farmers Association, 'NSW Farmers Association in focus', *The Land*, 17 July 2003.
59 Pritchard & McManus, *Land of Discontent*, p. 9.
60 Sydney Organising Committee for the Olympic Games, *Opening ceremony of the games of the XXVII Olympiad in Sydney*, Ultimo, 2000, p. 23.
61 Plumwood, 'Decolonising relationships with nature', pp. 10–11.
62 Wendell Berry, 'The whole horse: the preservation of the agrarian mind', in Andrew Kimbrell (ed.), *Fatal harvest: the tragedy of industrial agriculture*, Island Press, Washington, 2002, p. 11.
63 Les A Murray, *The vernacular republic: poems 1961–1983*, Angus & Robertson, North Ryde, 1988, p. 104.
64 Jimmy Little, 'Surely God is a lover', *Resonate*, Festival Mushroom Records, 2000, lyrics by Paul Kelly & John Shaw Neilson.
65 'Wallendbeen Cemetery', *Cootamundra Herald*, 26 November 1884.
66 Roderick Frazier Nash, *The rights of nature: a history of environmental ethics*, Primavera Press, Leichhardt, 1990, p. 91.
67 Carolyn Merchant, 'Reinventing Eden: western culture as a recovery narrative', in William Cronon (ed.), *Uncommon ground: towards reinventing nature*, WW Norton & Co., New York, 1995.
68 Vance Palmer, *National portraits*, Angus & Robertson, Sydney, 1940, pp. 197–8.
69 FJ Kendall, 'H.V. McKay', Grain Handling Authority of NSW, *Bulk Wheat Year Book*, 1984, p. 67.
70 Lynn White Jr, 'The historical roots of our ecologic crisis', *Science*, vol. 155, 1967, pp. 1203–7.
71 Peter Hay, *Main currents in western environmental thought*, UNSW Press, Sydney, 2002, pp. 100–6.
72 John Locke, *Two treatises of government*, Cambridge University Press, Cambridge, 1960 [1690], p. 309.
73 Society for Promoting Christian Knowledge, *The book of common prayer and administration of the sacraments*, Cambridge University Press, London, no date, p. 268.

Silence

1 Tucker, *If everyone cared*, p. 31.
2 'Old Aboriginal ill', *Sydney Morning Herald*, 14 March 1928.
3 'Marvellous is ill', *Yass Courier*, 12 March 1928; and 'Obituary', *Cootamundra Daily Herald*, 28 March 1928.
4 *Cootamundra Daily Herald*, 28 March 1928.

5   Gormly, 'Early days in this district'.
6   Gilmore, *Old days: old ways*, p. 138.
7   Read, *A hundred years war*, p. 64.
8   Tucker, *If everyone cared*, p. 91.
9   William Least Heat-Moon, *PrairyErth (a deep map)*, Houghton Mifflin, Boston, 1991, p. 243.
10 WEH Stanner, quoted in Tom Griffiths, *Hunters and collectors: the antiquarian imagination in Australia*, Cambridge University Press, Cambridge, 1996, p. 4.
11 Griffiths, *Hunters and collectors*, p. 4.
12 Merchant, *Reinventing Eden*, p. 227.
13 Deborah Rose, 'Connecting with ecological futures', The National Humanities and Social Sciences Summit, 26–27 July 2001, Canberra.
14 Gilmore, *The passionate heart*, pp. 70–2.
15 Tim Low, *Wild food plants of Australia*, Angus & Robertson, Pymble, 1991.
16 'A tour through the pastoral district of the Bland'.
17 Joseph H Maiden, 'A plea. Our brush forests', *Sydney Morning Herald*, 6 June 1908.
18 Joseph H Maiden, *Some New South Wales plants worth cultivating for shade, ornamental, and other purposes*, Department of Agriculture, Sydney, 1896, p. 9.
19 *Cootamundra Herald*, 25 January 1896.
20 Gormly, 'Early days in this district'.
21 Read, *A hundred years war*, p. 35.
22 John B Gribble, 1882 diary, Australian Institute of Aboriginal and Torres Strait Islander Studies, MS 1514/1, Item 3.
23 'The Warangesda Mission', *Cootamundra Herald*, 4 February 1882.
24 Kabaila, *Wiradjuri places*, p. 59.
25 Pamela Lukin Watson, *Frontier lands and pioneer legends: how pastoralists gained Karuwali land*, Allen & Unwin, St Leonards, 1998, p. 49.
26 Gilmore, *More recollections*, p. 34.
27 Gilmore, *The passionate heart*, p. 68.
28 Iris Clayton & Alex Barlow, *Wiradjuri of the rivers and plains*, Heinemann Library, Melbourne, 1997, p. 2.
29 Kabaila, *Wiradjuri places*, p. 133.
30 Clayton & Barlow, *Wiradjuri of the rivers and plains*, p. 7.
31 Kabaila, *Wiradjuri places*, p. 94.
32 Clayton & Barlow, *Wiradjuri of the rivers and plains*, p. 9.
33 Kabaila, *Wiradjuri places*, p. 68.
34 Read, *A hundred years war*, p. 35.
35 'NSW faces Mabo-style claim: Aborigines seeking native title over one-third of state', *Sydney Morning Herald*, 4 June 1993.
36 'Mabo grabs carry threat of backlash', *The Land*, 10 June 1993.
37 'Land bid violence warning', *Daily Advertiser*, 5 June 1993.
38 'Land claims are ludicrous: Nats', *Daily Advertiser*, 10 June 1993.
39 *Daily Advertiser*, 5 June 1993.
40 'Wiradjuri loses claim on huge area of NSW', *Sydney Morning Herald*, 24 December 1993.
41 'Reaction mixed to sign wording', *Cootamundra Herald*, 30 December 1998.

42 Gilmore, *Old days: old ways*, pp. 152–153.
43 Bill Gammage, *Australia under Aboriginal management*, 15th Barry Andrews Memorial Lecture, Australian Defence Force Academy, Canberra, 2002, p. 9.
44 Theodor W Adorno & Max Horkheimer, *Dialectic of enlightenment*, Verso Editions, London, 1979, p. 9.
45 'Fire as hearth or holocaust: it's our choice, says expert', *Sydney Morning Herald*, 6 October 2003.
46 Justin Murphy, 'Salinity – our silent disaster', accessed February 2004, <http://abc.net.au/science/slab/salinity/default.htm>; 'Bitter harvest: dryland salinity in Australia, *Sydney Morning Herald*, 3 December 1999; 'Dryland killer choking the land's lifeblood', *Advertiser*, 20 November 2001; 'Salinity: Qld's life or death decision', *Queensland Country Life*, 18 July 2002; and 'Salinity alarm bells ringing', *Queensland Country Life*, 23 May 2002.
47 Wendell Berry, quoted in Jackson, *Altars of unhewn stone*, p. 10.
48 George Myerson, *Ecology and the end of Postmodernity*, Icon Books, Cambridge, 2001, p. 41.
49 Berry, 'The whole horse', p. 10.
50 John Williams, 'Towards sustainable land management', *Food for healthy people and a healthy planet*, internet conference, September 2001, accessed February 2002, <http://www.natsoc.org.au>.
51 'New weapon to fight salinity', *Rural Weekly*, 25 January 2002.
52 'New CD-ROMs to help the war on salinity', *The Land*, online daily news, 1 October 2003, <http://theland.farmonline.com.au/news_daily.asp?ag_id=16788&s=4831>.
53 Australian Academy of Science, 'Monitoring the white death – soil salinity', accessed July 2003, <http://www.science.org.au/nova/032/032key.htm>.
54 'Howard turns his mind to establishing a legacy', *Australian*, 2 November 2002; and Jennifer Goldie, Kate McDonald & Fran Molloy, 'Feature article', accessed May 2002, <http://www.journalism.uts.edu.au/archive/salinity/feature.html>.
55 Joint statement by Warren Truss, Federal Minister for Agriculture, Fisheries and Forestry, David Kemp, Federal Minister for the Environment and Heritage, and Stephen Robertson, Queensland Minister for Natural Resources and Mines, AFFA02/164WTJ, 3 July 2002.
56 'Medicare-style 'environment levy' necessary to beat salinity, land clearing', accessed October 2001, <http://www.nccnsw.org.au/ncc/news/media/20011015_levy.html>.
57 Murray Darling Basin Commission, *How to encourage sustainable land use in dryland regions of the Murray-Darling basin*, accessed January 2003, <http://www.landmark.mdbc.gov.au>.
58 Jackson, *Altars of unhewn stone*, p. 15.
59 George Sessions, 'Ecological consciousness and paradigm change', in Michael Tobias (ed.), *Deep ecology*, Avant Books, San Diego, 1985, p. 33.
60 Ian Donges, 'Land management and asset security', *Connections*, Agribusiness Association of Australia, accessed March 2005, <http://www.agrifood.info/10pub_conn_Sm2001_6.htm
61 Worster, *The wealth of nature*, p. 50.
62 Worster, *The wealth of nature*, p. 110.

63 Berry, *Life is a miracle*, p. 19.
64 *The Land*, 4 November 2004.
65 *The Land*, 24 April 2003.
66 'Government acts on wheat virus', Department of Agriculture, Fisheries and Forestry media release, accessed July 2003, <http://www.affa.gov.au/ministers/truss/releases/03/03098wt.html>.
67 'Frost damage bill feared to top the $100 million mark', *Junee Southern Cross*, 29 November 2001.
68 'Cold blow for local farmers', *Southern Weekly Magazine*, 5 November 2001.
69 'Frost damage highlights relief needs', *The Rural*, 30 November 2001.
70 'Association branch president working to aid members hurt by the late frosts in October', *Junee Southern Cross*, 29 November 2001.
71 'The influence of an agricultural life', *Agricultural Gazette of New South Wales*, 1 February 1924, p. 103.
72 'British millers' requirements in wheat', *Agricultural Gazette of New South Wales*, volume IX, July to December 1898, p. 750.
73 AR Callaghan & AJ Millington, *The wheat industry in Australia*, Angus & Robertson, Sydney, 1956, pp. 407–408.
74 CJ Dennis, 'Wheat', in *Backblock ballads and later verses*, Angus & Robertson, Sydney, 1918, pp. 23–26.
75 Andrew Barton Paterson, *The man from snowy river and other verses*, Angus & Robertson, North Ryde, 1987 [1961], pp. 181–182.
76 Eric J Underwood, 'Man land and food', 1967 Farrer Memorial Oration, *Agricultural Gazette of New South Wales*, vol. 78, part 5, May 1967.
77 Ian Macdonald, 'Agriculture on the move', *NSW Agriculture Today*, 28 August 2003.
78 'Novartis: new skills in protecting the world's crops', National Farmers' Federation, *Australian agriculture*, Melbourne, 1997, pp. 8–9.
79 'They have banished hunger', *Agricultural Gazette of New South Wales*, February 1955, p. 57.
80 Julian Cribb, 'Agriculture's new role in the human destiny', Agricultural Science, vol. 11, no. 4, 1998, p. 23.
81 Incitec Fertilizers, *Fertilizer News: Grain*, Southern edition, autumn 2001, p. 11.

Revolt

1 Gilmore, *Old days: old ways*, p. 118.
2 Sally McNicol & Dianne Hosking, 'Wiradjuri', in Nick Thieberger & William McGregor (eds), *Macquarie Aboriginal words*, Macquarie University, 1994, p. 86.
3 Tucker, *If everyone cared*, p. 52.
4 Andrew Barton Paterson, *Illalong children*, Weldon Publishing, Sydney, 1989, pp. 41–2.
5 George Crooks, letter dated 2 November 1992, held by Ian Thompson, Temora.

6 'Bushland barometers', *'The Land' Farm & Station Annual*, 14 July 1937.
7 Keith C McKeown, *The land of the Byamee: Australian wild life in legend and fact*, Angus & Robertson, Sydney, 1938, p. 147.
8 NSW National Parks and Wildlife Service, 'Bush stone-curlew', Threatened species information sheet, 1999.
9 Barellan PA & I Society, *Early days in Barellan and district*, Barellan, 1975.
10 Clune, *Last of the Australian explorers*, p. 36.
11 HC Stening, 'Discing stubble land before ploughing', *Agricultural Gazette of New South Wales*, 2 January 1914, p. 27.
12 PG Gilder (ed.), *The farmers' handbook*, Department of Agriculture, Sydney, 1918, pp. 251–2; and Sutton, *Wheat-growing in New South Wales*, p. 16.
13 FJ Meurant, 'Fallowing and rotation of crops', *Agricultural Gazette of New South Wales*, 2 January 1915, p. 73.
14 Max Leitch, *Where the red gums are growing*, Oxford Print, Wagga Wagga, 1985.
15 Wantabadgery station diaries, Mitchell Library, A3320–A3338, 27 July 1879, 13 March 1886, 22 April 1886, 27 July 1886.
16 Department of Agriculture, inspection report, 15 February 1946, Wantabadgery West estate file, State Records NSW, 10/26077.
17 ES Clayton, 'The control of soil erosion on wheat lands', *Agricultural Gazette of New South Wales*, November 1931.
18 'Vanishing farmlands: trademarks of a reckless agriculture', *'The Land' Farm & Station Annual*, 14 July 1937.
19 Roland Breckwoldt, *The dirt doctors: a jubilee history of the Soil Conservation Service of NSW*, Soil Conservation Service of NSW, Sydney, 1988, p. 54.
20 Barellan PA & I Society, *Early days in Barellan*, p. 161.
21 Musgrave, *The wayback*, p. 27.
22 *Junee Southern Cross*, 19 December 1902.
23 Christine Jones, 'Native perennial pastures ignored in recharge–discharge model', *Australian Farm Journal*, January 2001, pp. 58–61.
24 Sherry Morris (ed.), *Junee: speaking of the past*, vol. 1, Junee Shire Arts Council, Junee, 1997, p. 6.
25 George Seymour, 'My early days', Wagga Wagga and District Historical Society newsletter, no. 294, June–July 1995, p. 4.
26 EDH, '"Old" Junee', article held by the Junee Public Library from an edition of an unidentified newspaper, 22 August 1932.
27 Barellan PA & I Society, *Early days in Barellan*, p. 87.
28 'The Great Bush Fires', *Wagga Advertiser*, 5 January 1905.
29 Bill Belling & Fay Belling, *Tarcutta stories*, 2nd edn, privately published, Tarcutta, 1991, p. 48.
30 Peter L Smith, Brian Wilson, Chris Nadolny, & Des Lang, *The ecological role of the native vegetation of New South Wales*, Department of Land and Water Conservation, Sydney, 2000, p. 26.
31 Greg Martin, 'Understanding soil building processes', *Stipa Native Grasses Newsletter*, no. 16, autumn 2001.
32 Barellan PA & I Society, *Early days in Barellan*, p. 263.
33 Barellan PA & I Society, *Early days in Barellan*, p. 10.
34 Rob Webster, *Bygoo and beyond*, Halstead Press, Sydney, 1956, p. 115.

35. Webster, *Bygoo and beyond*, p. 101.
36. Nigel Clark, 'Wild life: ferality and the frontier with chaos', in Klaus Neumann, Nicholas Thomas, & Hilary Ericksen (eds), *Quicksands: foundational histories in Australia & Aotearoa New Zealand*, UNSW Press, Sydney, 1999, p. 145.
37. Richard Littlejohn, *Early Murrumburrah historical notes*, Harden–Murrumburrah Historical Society, Harden–Murrumburrah, no date, p. 5.
38. 'Soil acidification: an insidious soil degradation issue', National Land and Water Resources Audit 2001, Environment Australia, accessed December 2002, <http://audit.ea.gov.au/ANRA/agriculture/docs/national/Agriculture_Soil_Deg.html>.
39. Swan, *A history of Wagga Wagga*, p. 137.
40. John Ross, inspection reports, Retreat Station, 1894 and 1896, ANU Noel Butlin Archives, 2/306/56.
41. Richard H Cambage, 'Notes on the botany of the interior of New South Wales', *Proceedings of the Linnean Society of New South Wales*, vol. 27, p. 194.
42. Richard Manning, *Against the grain: how agriculture has hijacked civilization*, North Point Press, New York, 2004, p. 28.
43. 'Post-storm clean up', *Cootamundra Herald*, 7 April 2003.
44. *The Land*, 5 December 1963.
45. Grain Elevators Board of New South Wales, *Bulk Wheat*, vol. 8, October 1974, p. 2.
46. Grain Elevators Board of New South Wales, *Bulk Wheat*, vol. 10, October 1976, p. 2.
47. Scott, *Seeing like a state*, p. 287.
48. Wellcome Australia Limited, 'Will smart insects inherit the earth?', in Grain Handling Authority of NSW, *Bulk Wheat*, 1985, p. 62.
49. 'Herbicides losing wild oat "punch"', *The Land*, 4 October 2001.
50. Stephen B Powles, 'Evolution in action: genes endowing multiple herbicide resistance in *Lolium rigidum*', paper delivered at the *XIX International Congress of Genetics*, Melbourne, July 2003, accessed September 2003, <http://www.geneticscongress2003.com/index.php>.
51. William Leiss, 'The domination of nature', in Carolyn Merchant (ed.), *Key concepts in critical theory: ecology*, Humanity Books, New York, 1994, p. 61.
52. Shiva, *The violence of the green revolution*, p. 21.
53. Carolyn Merchant, 'Introduction', in Merchant, *Key concepts in critical theory*, p. 4.
54. Gilmore, *The passionate heart*, p. 191.
55. Paul Carter, 'Turning the tables – or, grounding post-colonialism', in Kate Darian-Smith, Liz Gunner & Sarah Nuttall (eds), *Text, theory, space: land, literature and history in South Africa and Australia*, Routledge, London, 1996, p. 30.
56. HD Evans & NC Proud (eds), *Back to Junee*, Back to Junee Committee, Junee, 1947.
57. Morris, *Junee: speaking of the past*, p. 103.
58. SL Macindoe & C Walkden Brown, *Wheat breeding and varieties in Australia*, New South Wales Department of Agriculture, Sydney, 1968, p. 91.
59. FE Stanton (ed.), *Bust, boom and bust: some reminiscences of wheat and*

wheat breeding in Australia, Department of Agriculture, Sydney, 1984, pp. 106–107; and FJ McRae, *Winter crop variety sowing guide 2001*, NSW Agriculture, Orange, 2001, p. 11.

60 'Oat variety sowing guide for 2003', Primary Industries and Resources SA, fact sheet 9/86/03.

61 Andrew Barton Paterson, *'Banjo' Paterson's Song of the Wheat: a reproduction of an illuminated manuscript by Gordon Nicol*, National Library of Australia, Canberra, 1988 [1925].

62 Joseph W Dwyer, quoted in Sylvia Walsh, *Dr. Joseph Wilfred Dwyer D.D., first bishop of the Diocese of Wagga Wagga*, privately published, Wagga Wagga, 1978, p. 16.

63 Fleur Stelling (ed.), *South west slopes revegetation guide*, Murray Catchment Management Committee and Department of Land and Water Conservation, Albury, 1998, p. 240.

64 GM Cunningham, WE Mulham, PL Milthorpe & JH Leigh, *Plants of western New South Wales*, Soil Conservation Service of New South Wales, Sydney, 1981, p. 528.

65 Joseph W Dwyer, 'A floral survey of the south western slopes of N.S.W. round about Temora and Barmedman', *Australian Naturalist*, vol. 4, July 1921.

66 Laurel Thompson, 'Berrin Berran (the Place of the Grey Box Trees)', in *Temora yesterday and today 1880–1980*, Temora Centenary Historical Book Committee, Temora, 1980, p. 17.

67 Walsh, *Dr. Joseph Wilfred Dwyer D.D.*, p. 16.

68 Steele, *Temora's jubilee souvenir*.

69 Joseph W Dwyer, 'Wattles', *Australian Naturalist*, vol. 9, June 1933, p. 12.

70 John Rudder & Stan Grant, Wiradjuri language classes, 2001 and 2002, Charnwood, Canberra.

71 Fred R Myers, *Pintupi country, Pintupi self: sentiment, place, and politics among Western Desert Aborigines*, Smithsonian Institutional Press, Washington, 1986, p. 48.

Regeneration

1 Joseph H Maiden, *The useful native plants of Australia (including Tasmania)*, Compendium, Melbourne, 1975 [1889], p. 15.

2 Bill Gammage & Peter Spearritt, *Australians 1838*, Fairfax, Syme & Weldon, Sydney, 1987, p. 118.

3 George Day & Julie Follett (eds), *Aboriginal arts in Transition*, NSW Department of Eduction, Wagga Wagga, 1988, p. 49.

4 Ardrossan historical timeline', accessed May 2003, <http://www.ayrshire-roots.com/Towns/Ardrossan/Ardrossan%20History.htm>.

5 Berry, *What are people for?* pp. 206–209.

index

Aboriginal Girls Training Home 114, 164, 173, 176, 177
Aboriginal law 17–18
Aboriginal people
   children 114–15, 164–5, 173–4, 176–8
   identity 105–6, 112–14, 116–19, 163–4, 173–4, 177–81
   see also Gudhamangdhuray people; Wiradjuri people
Aborigines Protection Board 12, 114, 164, 173, 176, 177
Abram, David 106
Adorno, Theodor 184
agricultural development 7, 42–56, 64–9, 74–87, 102, 103–7, 147, 158, 168, 172, 190–7, 220–2, 224
   see also farming and the environment
agricultural technologies 54–6
Aitkin, Don 148
Albury 63
Allowrie 72
ants 27
Ardrossan 238–42
Avcare 122

Baldry, Alex 45
Bardwell, Maria 10
Barellan 202
Baylis, James 21
Beckham, Edgar 22, 30
Bell, John 115
Bell, Melinda (nee McGuiness) 110, 112, 113, 114–20, 163

Belling, Bill and Fay 12
Bennett, George 23
Berry, Wendell 8, 87, 152, 185, 186, 192, 248–9
Bethungra 20, 37, 38, 181
Bethungra Park 38–9, 40
Billygoat Hill 107, 108, 109, 110, 111
Bingham, Henry 18–19
biological diversity 6–7
Birrego 242, 244, 246
Blainey, Geoffrey 148
Blakely, William 224
Bland Creek 40, 41–2, 43
Borlaug, Norman 124
Bourke, Lisa 148
Bradman, Donald 175–6
Brae Farm 72
Breakwell, Ernest 48
Brody, Hugh 121
Brown, Ted 212–14
Brungle 105, 112, 113, 114, 116, 178
Buckargingah Creek 52, 54, 55
Bulger, Vince 113
Burrangong Creek 32
Burrangong pastoral station 17, 58

Caldwell, Steel 31–2
Cambage, Richard 217
Cameron, Donald and Mary 24, 27, 28–9, 57
Capra, Fritjof 7
Carson, Rachel 123
Carter, David 193
Carter, Paul 222

Cawley, Robert 13
Chakrabarty, Dipesh 99
Chinese thistle 46
chemicals 50–1, 122, 123, 126, , 218–19, 220
chocolate lilies 14–15, 26, 70, 133, 167, 224, 256
Christianity 157–60
Clark, Nigel 211
Clayton, Iris 176, 177
Clayton, Sam 206–7
Clements, Theresa 165
Clune, Frank 30, 204
Coe, Paul 178, 179
colonisation 6, 14, 17, 57–9, 83, 89–90, 104–5, 108, 126, 128–9, 132–4, 139, 158, 180, 184, 210, 222
Congou Creek 31
Congou Hill 29–30, 31
Connaughtmans Creek 69–70, 72
Cobborn Jackie 58
Conway, Jill Ker 8
Cootamundra 30–1, 36, 62, 63, 64–5, 73–5, 86, 93–5, 97–101, 107–20, 134, 135–6, 140–1, 164–5, 171–81, 227, 228
Cootamundra Development Corporation 4, 141, 175
Corby, William 65
Cosby, Henry 19
Cowley, Frank 38
Cribb, Julian 196
Crooks, George 201
crop monocultures 7
Crowe, Alan 172
Cullinga 72
culture, rural and urban 147–54, 261
curlew 201–2, 222
Currawong 212

Davison, Bruce 43
Davitt, Michael 96
de Chair, Sir Dudley and Lady 73
Deepwater station 21–2, 25, 28
Dennis, CJ 194
Descartes, Rene 47
Devlin, James 21–2
Donald, Colin 39, 43, 103

Donges, Ian 191
Donnelly, Edward 216
Druce, Peter 218
Dwyer, Joseph 223, 224–5
dryland salinity 8, 56, 80, 82–3, 184–90, 220
Dudal Swamp 55–6

eastern snake-necked turtle 4, 155, 173
ecological
    change 9, 208, 209, 210
    disorders 7–8, 9, 51–2, 81–7, 104, 126–7, 172, 188, 203–9, 220
    diversity 133–4, 203
    stability 217
ecological sustainability 81, 84–7, 125, 245, 254–6
Eldershaw, Marjorie 146
emus 28, 169
environment and farming 121–7, 190–7, 220
erosion 7, 145–6, 205–7, 217
Eulomo station 20

Fackenheim, Emil 14
farm business risks 76–7
farming and the environment 121–7, 190–7, 220
    see also agricultural development
farmland management 220
Farrer, William 35–6
Faulks, Ross 75
fauna 4, 5, 25–6, 28, 50–1, 155, 169, 173, 201–2, 212–13, 222–3, 229–30, 231
fire 26, 33, 34, 88, 183, 181–4, 209–10, 256–7
Fisher, Elizabeth 211
Flanagan, Richard 138
flora 5–6, 14–15, 26, 29–30, 46, 70, 71, 72–3, 110, 127, 133, 145, 156, 167–71, 190, 197–8, 207, 214, 216–17, 222–3, 224–5, 228–9, 230, 231–8, 242–4, 247–8, 251, 255, 256–7
forests 33–5, 214
Fosbery, Leonard 34
Fox, Paul 129–30

Frankel, Otto 48
Franklin, Lady Jane 12–13

Galong 30
Gammage, Bill 21, 59–60, 183
Ganmain station 21–2, 25, 28
gardens, domestic 127–35
Gilbert, Kevin 100, 105
Gilmore, Mary 5, 19, 23, 24, 25, 26–7, 29, 37, 42, 57, 88–9, 144, 164, 168–9, 174, 175, 182, 184, 200, 202, 204, 221
Gilpin, William 137, 139
Glanville, Bill 115
Glanville, Bob 5, 110–12, 113, 115, 117–19, 163, 225
global markets 67–9, 76–7, 78, 83–4, 86–7
Godman, Bill 198
golden moth orchids 198
Goodall, Heather 6
Gormly, James 19, 21, 31, 34, 59, 111, 113, 163–4, 172
Gould, William Buelow 138
Gow, George 202–3, 204, 208, 209, 211
Grant, Cec 179
Grant, Stan 178
grassy woodlands 6, 7, 8, 33–4, 213, 222, 245, 255, 257
Gray, George 5
Gray, Ian 150
Greening Australia 214
Gribble, John 95, 173
Griffiths, Tom 8, 97, 167
Grogan 40
groundlarks 29
Gudhamangdhuray people 5, 108, 109, 116, 119, 172, 226
Gwynne, Richard 209

Hallam, Jack 75
Hammond, Thomas 209
Hansen, Alec 32–3, 34, 115, 207, 216, 228–31
Harden 61, 78–80
Hardies Reserve 70–2
Harris, Mark 81
Hawkins, Sara 42

Henty 52–5
herbicides 123, 126, 219, 254
Holloway, John 210
Horkheimer, Max 184, 219, 220
horses, wild 144
Houlaghans Creek 205
Howitt, Alfred 5, 172
Hull, Kay 124
human culture and animal nature 5
Hurley, John 31, 109, 113
Hurley's Springs 109

industrial production 124–5
industrial technologies 132–3, 227
Ingram, Ossie 236, 237
insecticides 122, 126, 218–19, 254

Jackson, Wes 86, 188
Jenkins, Frank 21, 61
Jindalee State Forest 32–3
Jones, Christine 208
Jugiong Creek 3, 79, 109
Junee 22, 63, 74, 76, 79, 111, 159, 160, 181, 205, 206, 208, 209, 222, 252

Kabaila, Peter 114, 174
Kavanagh, Harry 22
Keniston, Kenneth 8
'king plates' 18

Lackey, John 62
Lamp, Charles 216
land
    burning practices 33, 34
    clearing 35–40
    closer settlement legislation 89, 90
    for cropping 42–52
    degradation 150
    squatting 17–22, 58, 89–93
    wildlife reserve 28–9
Landcare 51, 52, 81–2, 85–6, 149, 189, 246
Landy, Lynne 130
Larmer, James 42
Last, Sally 198
Lawrence, Geoffrey 78, 150
Leary, Joseph 63

Leitch, Max 205
Leopold, Aldo 79, 120, 130
Linn, Rob 51
Little, Jimmy 155–6
Littlejohn, Richard 212
Locke, John 158
Lockie, Stewart 148
Lumholtz, Carl 94

McCubbin, Frederick 147
Macdonald, Ian 54, 85, 196
McGuiness, Arthur 113–14, 115, 117, 225–6
McGuiness, Eric 116
McGuiness, Iris 114, 115
McGuiness, Marie 112
McGuiness, Melinda *see* Bell, Melinda
McGuiness, Minnie 115
Mackay, Alexander 30, 31, 41, 90
Mackay, Donald 30, 204
McKay, HV 53, 102, 157
Mackay, Kenneth 31, 41, 90–2
McKeown, Keith 202
MacLennan, Caroline 37
McNamara, Veronica 31–2
Maiden, Joseph 170–1, 224, 236
Manning, Joe 231
Marshall, William 101
Maslin, Ron 101–2
Mate, Thomas Hodges 10–11, 12–13, 14
Mates Gully Creek 14
Mathews, Freya 67, 122
Mathews, Janet 114, 225
Marx, Leo 8
Merchant, Carolyn 168, 220–1
Mill, John Stuart 67
Mimosa station 36, 144
Mitchell, Thomas 20, 33–4
Mitta Mitta 76, 77, 256, 259, 260
Mofflin, Wilcox 50
Moombooldool station 202, 210
Morangarell district 40
Morangarell station 31, 41
Morgan, James 176
Morrow, James 71, 72
Morton, Mary 31, 223
Muecke, Stephen 143
Murray, Les 152–4

Murray-Darling Basin 124, 187
Murrumbidgee River 59–61
Murrumbidgee River catchment 3
Murrumburrah 61
Musgrave, Sarah 17, 30, 58, 208
Muttama Creek 3, 4, 112, 174–5
Muttama station 59
Myerson, George 185

Nancarrow, Justin 13
Narrandera 20, 63
Nash, Roderick 156
native plants *see* flora
natural resource management 8
Ngarrangdhuray people 21–2
Nicol, Gordon 223
Noble, John 163, 164, 165, 166
Noble, Robert 44

Oakley 252, 258–9
old man weed 110
Open Garden Scheme 129–30
Oxley, John 139

Parkan Pregan lagoon 25, 27–8
Parramore, Olive 12
Paterson, Banjo 194–5, 201, 223
pest control 50–1, 214
Piccoli, Adrian 123
Pinchgut Creek 144, 145–6, 159, 161, 214, 216
Plumwood, Val 120, 126, 130, 133
Podmore, Harry 13
Pregan Island 200–1
Pyne, Stephen 184

railways 52, 60, 61–5, 70, 109, 172
Read, Peter 18
Retreat 143–4, 145, 159–60, 161–2, 214, 216, 240
Rickard, John 147
ringbarking trees 35–8, 72, 206, 213, 216, 231–2, 240
riverboats 59–61, 63
Roberts, James 212
Roberts, Richard 61
Roberts, Tom 147
Roberston, John 24

Robinson, Michael 93
The Rock nature reserve 120–2
Rolles, Allan and Paul 182
Rolls, Eric 33
Rose, Deborah 14
Rosevale 212–13
Ross, John 216
rural culture 147–54, 261
Ryan , Ned 30, 31
Ryan, Simon 138, 139
Ryan, Terry 178

Said, Edward 108
salinity 8, 56, 80, 82–3, 184–90, 220
SalinityBiz 83
sanctuary regulation 5–6, 25–8, 88–9, 172–3, 200–1
Sanitarium Health Food Company 66, 67
Sawyer, Mat 20
Scott, James 7, 140, 218
Seed, Patricia 128
Sessions, George 191
Seymore, George 209
Sheehan, Norm 17
Shiva, Vandana 76, 104, 124, 125
sifton bush 216–17, 220
Smith, David 46, 48, 86, 104
Smith, Irwin 71, 72
Smith, Richard Bowyer 102
Smith, Tim 13
social and spiritual concerns 6
soil acidification 80, 213, 215, 219–20, 254
soil conservation 206–7
squatters 17–22, 31, 57, 58, 89–93
Stanner, WEH 167
Stanyer, Charlie 201, 209
Steele, Watson 224
Stockinbingal 16, 29, 30, 31, 41, 170, 197–8
Strong, Graham 244–8, 249–50
Strong, John 193
Sturt, Charles 18, 138–9
Sturt, Evelyn 88
Sutton, George 35, 36
swamplands 5, 6, 7, 8, 112, 145, 172, 175

swans 25–6
Sweeny, Francis 71, 72

Taaffe, Francis 59, 163
Tacey, David 47
Tang 12
Tarcutta 10, 12–13
Tarcutta Creek 13
Taylor, Headlie Shipard 53, 102
Temora 20, 63, 65, 95, 99, 100–2
Temora Demonstration Farm 45
Thompson, Edward Deas 19
Thompson, Ian 37, 38
Thompson, Laurel 224
Thompson, Paul 78
tractors 51
trading, global 67–9, 76–7, 78, 83–4, 86–7
Truss, Warren 86
Tucker, Margaret 114, 163, 164–5, 200
turtles, freshwater 4, 5
Turner, Ned 12

Underwood, Eric 47, 48–9, 195

Waddy, John 98
Wagga Wagga 60, 64, 96–7, 200
Wallendbeen 29, 30, 31, 70, 71–2, 90, 92, 156, 167–8, 190–1
Wantabadgery station 206
Ward, Russel 148
Watson, James 61
Watson, Pamela Lukin 174
Watt, Robert 43, 132
Webster, Rob 35, 65
Wedge, HJ 134
weeds 46, 169–70, 205, 207, 216–17
West, Ethel 41
Westmacott, Charles 38
Weston, Fred 162–3, 165, 166
wheat 74–7, 78–9, 132, 157, 168, 196
Whitaker, Hugh and Catherine 252, 259
Whitaker, Joe 260
Whitaker, Owen 40, 76–7, 83, 251, 252–61
White, Chum 12

White, James 17–18, 58
White, John 58, 179
White, Lynn 158
Whitton, Lesley 177
Williams, John 20, 124, 186
Williams, Roly 12
Wiradjuri people 4, 5–7, 11–12, 16–17, 22–3, 24–5, 31, 56, 109–13
   Aboriginal identity 105–6, 112–14, 116–19, 163–4, 173–4, 177–81, 225–6, 227
   conflict with settlers 16–24, 30–2, 57–8, 93–5, 100, 101, 113, 168, 180
   hunting and gathering 23, 24–5, 33, 73, 88, 169–70, 198–9, 208
   land burning practices 33, 34, 88, 183, 256–7
   language 16
   law 17, 246
   population 22
   sanctuary regulation 5–6, 25–8, 88–9, 172–3, 200–1
   see also Aboriginal people
Woodside, Dedee 123
Worster, Donald 8, 126, 192
Wright, Judith 131

Yarranjerry station 211
Yonco station 20
Young 17, 58, 208

OPPOSITE
Cutting chaff, Harden district, about 1900.
Courtesy Harden-Murrumburrah Museum.

Marinna railway silos, Junee district, 2001.

Carting wheat to the railway at Harden from Lulworth, a farm owned by the Kendall family, about 1900. Courtesy Harden-Murrumburrah Museum.

BACK COVER
Muttama Creek waterhole, Cootamundra, October 2003.

Mural by Dale Huddleston beneath the Yass River bridge, Yass, 2005.

Chocolate lily flower, Oakvale, Narrandera district, October 2004. Photograph by Rosie Smith.